CW00631760

An Extraordinary Land

An Extraordinary Land

DISCOVERIES AND MYSTERIES FROM WILD NEW ZEALAND

Written by Peter Hayden
Photographs by Rod Morris

HarperCollins*Publishers*

To the men and women, boys and girls who love and look after this extraordinary land

The authors and publishers wish to thank and acknowledge the following individuals and organisations who have provided photographs: Marc Backwell/ *New Zealand Geographic*, page 128 (upper); Backyard Kiwi, page 179 (lower); Peter Hayden, page 103 (lower); Hocken Library, page 185 (lower); Norm Judd/ DOC, page 153 (both photos); David Kleinert, page 23; Don Merton, pages 61 (lower left), 121 (upper), 158–159 (all photos); Merton family, page 194; National Maritime Museum, page 28 (upper left); Orbell family, page 112; Tom Simpson/ *New Zealand Geographic*, page 133 (both photos); Mike Soper, pages 122 (lower), 157; Southern Discoveries, page 195 (lower); Bruce Thomas, pages 160 (upper), 160–161 (lower); Pete Tyree/DOC, pages 148–149, 152 (lower), 154 (upper left), 155; USGS/NASA, pages 6–7; Wet Tropics, page 138 (centre). All other photographs are by Rod Morris, www.rodmorris.co.nz.

HarperCollins*Publishers*

First published in 2013
by HarperCollins*Publishers (New Zealand) Limited*
PO Box 1, Shortland Street, Auckland 1140

Photos copyright © Rod Morris (except as indicated above) 2013
Text copyright © Peter Hayden 2013

Rod Morris and Peter Hayden assert the moral right to be identified as the authors of this work.

All rights reserved. No part of this publication may be reproduced, stored in a retrieval system or transmitted in any form or by any means, electronic, mechanical, photocopying, recording or otherwise, without the prior written permission of the publishers.

HarperCollins*Publishers*
31 View Road, Glenfield, Auckland 0627, New Zealand
Level 13, 201 Elizabeth Street, Sydney, NSW 2000, Australia
A 53, Sector 57, Noida, UP, India
77–85 Fulham Palace Road, London W6 8JB, United Kingdom
2 Bloor Street East, 20th floor, Toronto, Ontario M4W 1A8, Canada
10 East 53rd Street, New York, NY 10022, USA

National Library of New Zealand Cataloguing-in-Publication Data
Morris, Rod, 1951–
An extraordinary land : discoveries and mysteries from wild New Zealand /
Rod Morris and Peter Hayden.
Includes index.
ISBN 978-1-86950-963-7
1. Natural history—New Zealand. 2. Endemic animals—New Zealand.
I. Hayden, Peter. II. Title.
508.93—dc 23

ISBN: 978 1 86950 963 7

Cover design by Darren Holt, HarperCollins Design Studio
Cover images by Rod Morris
Typesetting by IslandBridge
Colour reproduction by Graphic Print Group, South Australia

Printed by RR Donnelley, China

Contents

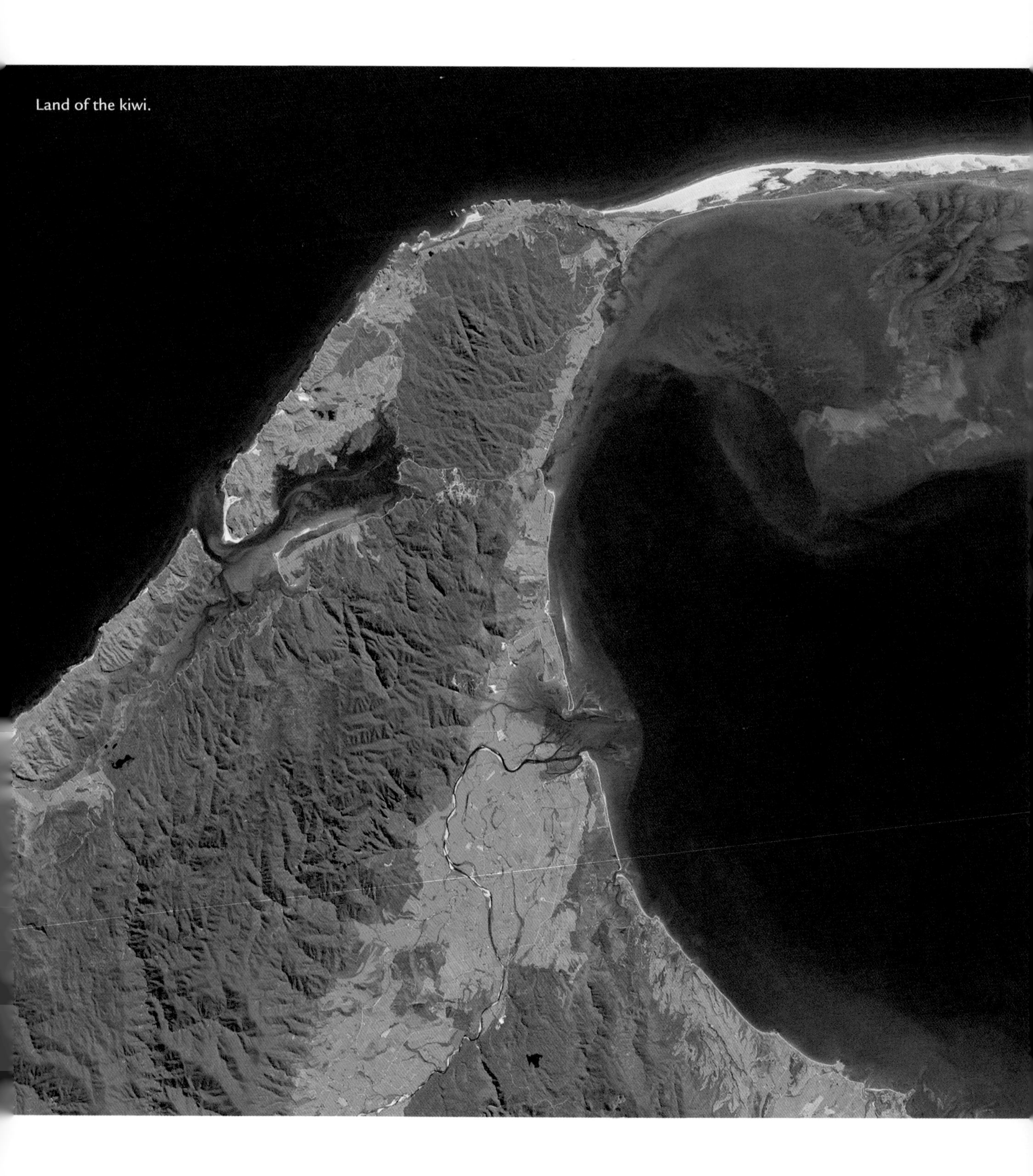
Land of the kiwi.

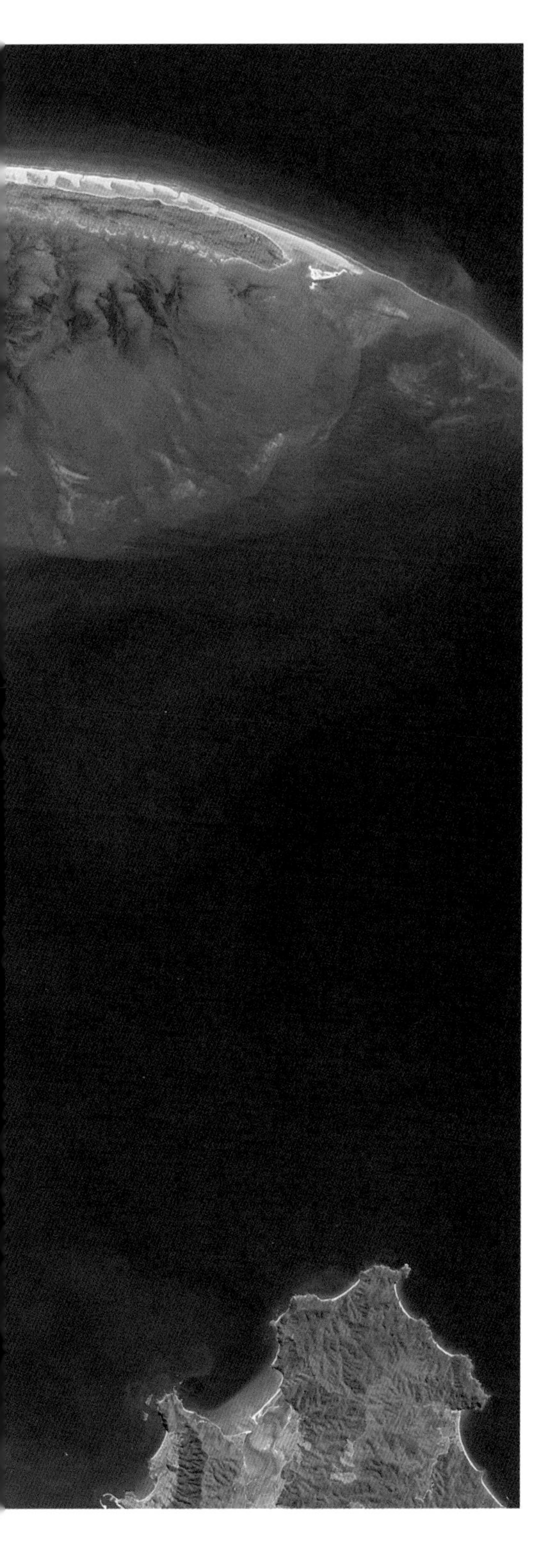

Introduction

New Zealand is an extraordinary land. It has variously been described as 'the closest thing to life on another planet', 'a lifeboat', 'Moa's Ark', and so forth. We live on a group of islands that aren't like the rest of the world. How did they get so strange?

The main reason is that for a very long time our islands remained an isolated outpost for life in the midst of a vast ocean. That isolation was a huge advantage. It turned New Zealand into a wild laboratory where evolution could conduct experiments that led to weird and wonderful outcomes like the kakapo, a flightless, nocturnal, overweight parrot, and the giant weta, an insect the size of a mouse. It was also a museum, a refuge where ancient creatures such as tuatara could exist long after their kind disappeared from the rest of the world.

All this is great, and many scientists from around the world are itching to come and meet our eccentric plants and animals. Yet many of us who live here have no idea what's so special about our wildlife and our place in the world. This book offers a chance to understand and celebrate what puts our wildlife in a class of its own. It's also a chance to explode a few myths, solve a few mysteries and get alongside those working hard to rescue species in trouble.

In fact, showing New Zealanders the efforts to save species in trouble is what first brought Rod Morris and me together. We worked together to craft words and pictures into documentary films about our wildlife in the *Wild South*. In the beginning they were sad stories, almost obituaries, as some of the birds, such as black robins, takahe and kokako, were close to extinction. Thankfully, most species that were on the brink are now surviving and moving away from danger as a result of a lot of dedicated effort from many people.

More recently, scientists are discovering surprising new facts about species we thought we knew well. The kiwi is a classic example, and is the subject of some fascinating new research.

In these chapters we have tried to bring you up to date, give you the state of play, or present a snapshot of what is happening and what is so special about New Zealand wildlife.

Mystery Solved

Some stories explore our animals' strange behaviour or unusual lifestyles. For instance, why do we have so many nocturnal birds in New Zealand when almost all European birds that live in this country are out and about during the day? In *Fear*, we find out why so many native birds persist in skulking about at night.

The modern science of genetics has been a game changer. It has overturned some firmly held ideas and in some cases rewritten stories of our animals and plants. Genetics has revealed, for instance, how New Zealand's pohutukawa became such great *Explorers* of the Pacific, their progeny even spreading beyond the equator. What makes these explorers so resourceful is the manner in which they travel vast distances and colonise extremely harsh locations.

Genetics also opens up an intriguing *Cold Case* and helps solve the mystery of what happened to our little blue penguins after the ice ages. It seems they do not like living in cold places.

Who Knew?

We call ourselves Kiwis, but scientists are showing us that we know almost nothing about our national icon. Who knew that it is a *Sensory Superstar* with one of the biggest brains, relative to body size, of any bird in the world? We're only just learning the truth, too, about our *Pollinators*. This is the age-old story of birds and bees, right? Wrong. In New Zealand the 'deed' is just as likely to be done by lizards, and, as has recently been discovered, native bats are also first-rate pollinators!

Bob enlightens us about one of our little-known animal groups. It was the late scientist Bob McDowall who first solved the mystery of where whitebait come from, proving that the great New Zealand delicacy is made up of juveniles of native freshwater fish called galaxiids. Throughout his 40-year career, he continued to make remarkable discoveries about our freshwater environment.

Modern wildlife science has led to many new discoveries. Studies in genetics, for instance, show that New Zealand pohutukawa have spread to distant islands of the Pacific.

We call ourselves Kiwis but behavioural scientists are revealing that we know almost nothing about our national icon.

Why are so many of our species supersized and why do we spend so much time and effort protecting them from the jaws of introduced predators, such as rats?

These native species, like the brightly coloured takahe with its ruby-red beak and head plate, may be giants, but they are gentle giants.

Some great discoveries have been made not by scientists but by a group that has seldom been acknowledged. I am talking about children and their inquisitive nature. In *Curious* we celebrate the under-12s, profiling their discoveries and their contributions to our knowledge of native wildlife.

That's Weird

Another part of the New Zealand story can be summarised in one word: weird. In *Giants* we find ourselves in a world of super-sized species and ask the question, why so many? We offer a fascinating theory as a possible answer. And what could be the subject of *Waddling Mouse or Boxing Bat*? This is the story of a mysterious lost world that once existed in New Zealand. We meet the explorers of this alien world whose ongoing discoveries continue to excite the international scientific community.

To the Rescue

Much of the work of preserving native species comes down to saving them from the jaws of introduced predators. *Playing God* is what we do when we cause ecological upheaval or ecological salvation. This is the story of one of the boldest attempts ever undertaken in New Zealand to eliminate a predatory pest and restore a habitat. And in *Kiwi Comeback* we see how people power is becoming a potent force for good in safeguarding the future of our iconic national bird.

Protecting endangered species and restoring habitats continues to underpin a lot of conservation effort in this country. We take a look through *A Window* and glimpse a hopeful future where communities and businesses work alongside the Department of Conservation to preserve and expand safe refuges for our native animals and plants.

This book has been a journey of delight and discovery, bringing Rod and me into contact with the native creatures that make up this extraordinary land and the wonderful people who work so hard to protect them and increase our understanding of them. We hope that you will also be delighted and make your own discoveries in your journey through these pages.

Chapter 1

Fear

I used to be afraid of the dark. As children, our imaginations go into overdrive as shadows take on fearful shapes, and natural sounds are magnified and distorted into the soundtrack of a nightmare creature as it approaches the bed. *It's going to get me!* But when day finally dawns, our fears evaporate as quickly as sunrise. We are transformed. We are invincible, untouchable and completely unafraid.

But, imagine the flip side of fear of the dark. Imagine being afraid of light. It seems ridiculous. What ghost or monster could possibly inhabit our sun-drenched land of colour, warmth and gentle breezes? You would be surprised.

Fear of the light is real in New Zealand, which seems strange, as this land is a very unscary place. There are no big predators. In fact, to face real danger out in the bush, apart from getting lost or swept away in rapids, or falling off a cliff, you have to work at it. You have to either fall into a stinging nettle bush or eat something poisonous like tutu or karaka berries. But it's a stretch to identify something really scary out there. In fact, New Zealand was such a benign place to live that its birds evolved along lines best described as eccentric. The kiwi is a bumbling hairball with a beak; the kakapo is like a cuddly Persian cat; wrens are powder puffs with legs; kokako are squirrels that sing like divas.

If there are no scary monsters around, we invent them. Take weta, for example. How many of us shriek when a weta blunders out of the firewood? The problem is that we regard weta as ugly creatures and we have demonised them, turning ugly into scary. Really, they are ancient, oversized grasshoppers, whose only defence is to raise their spiky legs above their head. Unnerving as this is, there are weta out there that exhibit more extreme reactions to danger.

High in the Kaikoura ranges and the mountains of North Canterbury the bluff weta ambles about, feeding on alpine plants. It looks oblivious to intruders, but if you create a shadow or touch it, its reaction is spectacular. Instantly it rolls into a ball, and with its legs protruding like spokes, it leaps into space, rolling downhill for hundreds, or even thousands, of metres until it comes to rest. Then, if it has survived, it unfurls and wanders off. Why does it react so? There is no danger up here that requires such an extreme response. The only possibility is that

Opposite
The Mercury Island giant tusked weta has a face any demon would be proud of. A carnivore that eats earthworms and grubs, it is at risk itself from predation by lizards, from which it gains some protection by living in an underground burrow.

The bluff weta is a harmless vegetarian that lives on exposed mountain faces. It is a great climber, moving one leg at a time making sure all five others maintain a firm grip on the rock. But if danger threatens, it adopts what appears to be a suicidal response, rolling downhill.

it is a reaction to great danger that no longer exists. Perhaps a large predatory lizard once prowled these mountains, but has become extinct. Whatever the danger, the bluff weta retains a hard-core, hard-wired defence.

If you have a good torch and a stroke of luck, then by quietly moving along stream edges in inland Poverty Bay you may come across the Raukumara tusked weta. But chances are you won't. It will likely have sensed you before you are aware of it and will have leaped into the stream, where it will stay, unmoving, undetected, until danger passes. Then it will climb out onto the bank and resume feeding. Like the bluff weta, it has a hard-wired response to some long-lost enemy. These days it is a handy defence against rats.

Many New Zealand creatures have instincts for survival that seem unnecessary, even peculiar. For instance, why are so many birds, reptiles and insects nocturnal? Why are some, like the kea, dull coloured, with bright colours hidden under their wings? Why do others, like kiwi, freeze when danger lurks? Why do saddlebacks dive for cover? What are they afraid of? Have they seen a ghost? As it turns out, that's fairly close to the truth.

As for the kakapo, the idea of a nocturnal parrot seems absurd, and almost unknown anywhere else in the world. Parrots are creatures of light and treetops. By skulking about

Opposite

We think of parrots as typically colourful, but kakapo plumage makes a highly effective camouflage that would be the envy of any army fighting in the jungle. The kakapo has also turned its back on normal parrot behaviour by becoming flightless and nocturnal.

Kea rely on colour for sexual display, but at rest hide their bright colours under their wings, presumably to escape the attentions of predators. Colour and camouflage are both important, but at different times.

The saddleback (left) is a member of the wattlebirds, a group that also includes the kokako (above) and the extinct huia. Wattlebirds are weak-winged and fly only short distances, remaining within the forest canopy, away from predators.

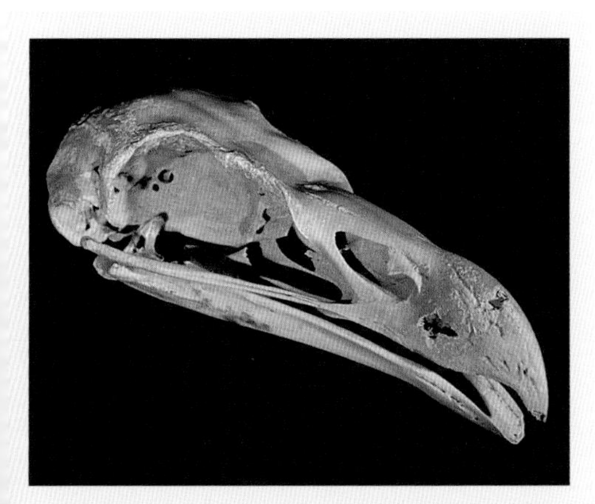

Most eagles of the world use their beaks and claws as major weapons. However, New Zealand's extinct Haast's eagle had claws the size of a tiger's. It was these formidable talons that inflicted most damage, as can be seen on this scarred moa bone. (See also page 20.)

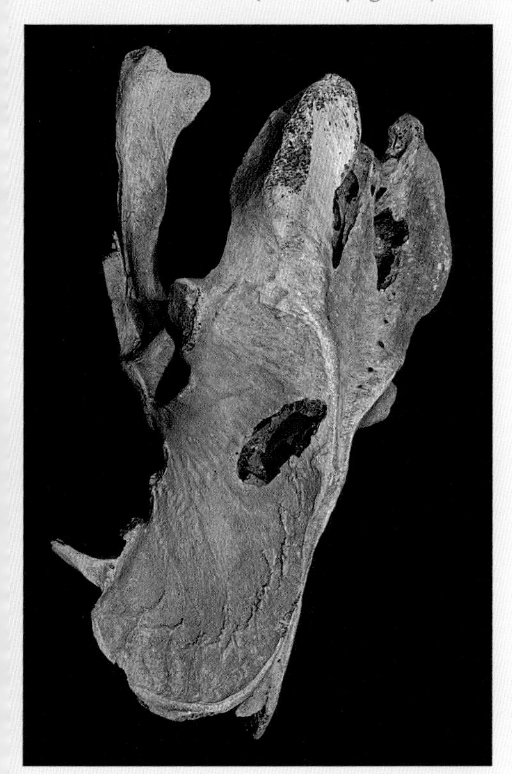

in the dark, what is the kakapo hiding from? And the kiwi, too, is another nocturnal bird. It even stops calling if the moon is bright. There is no obvious reason why kiwi should need to be so resolutely nocturnal, except to take advantage of prey. Worms are more active at night.

Even birds of the day are prone to the jitters. At Maungatautari, a 3400-hectare, predator-free mainland island south of Cambridge, manager Chris Smuts-Kennedy reports that when an RNZAF Skyhawk suddenly appeared overhead, banked and headed back to Ohakea air force base, a saddleback that had been feeding nearby dropped like a stone out of the canopy, crashing through the foliage to hide among the ferns. The plane was loud, but Chris considered the saddleback's reaction was over the top.

Could it be that the saddleback mistook the Skyhawk for an aerial attack by another 'sky hawk'— one that became extinct 600 years ago? Such a predator did exist. It was an eagle, the biggest, most ferocious eagle on the planet. It can truly be called New Zealand's scariest creature. In fact, there were two eagle-like predators in New Zealand. One was a harrier hawk, three times the size of the swamp harrier we see along roadsides and in open country. The other was a much bigger creature called Haast's eagle.

Julius von Haast, the first director of Canterbury Museum, found the first bones of this magnificent and terrifying bird. He called it *Harpagornis* from the Greek word *harpax*, meaning 'grappling hook'. It is aptly named. The talons are huge, like tiger claws, and the bones to which they are attached are much sturdier than those of any modern eagle. From bones and talons, Haast tried to imagine how and what this bird hunted. One hundred and forty years later we are still speculating, but we now have many more clues to work with, and this bird's remarkable story is emerging.

Complete skeletons have now been found. Using the circumference of the femur as a basis for calculation, it is estimated that females weighed over 15 kilograms; the skeletons also reveal that they had a wingspan of up to 3 metres, twice that of the largest living eagle, Steller's sea eagle of north-eastern Asia. But there the comparison ends.

An Australasian harrier is 'mantling' its prey. This is an aggressive or threat display to make it appear bigger to predators, and to keep other harrier hawks away from its freshly killed meal.

A comparison of the skull, breast, leg and claws of the much larger Eyles' harrier (left), which is now extinct, and the Australasian harrier (right) that we see today on roadsides and in open country.

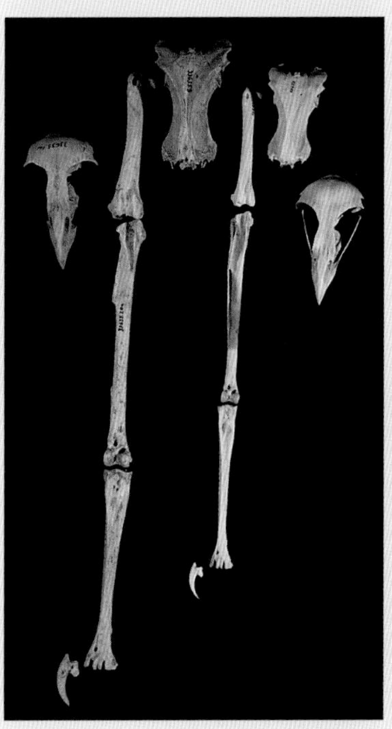

Moa remains have been found with puncture marks in their pelvic bones. Only *Harpagornis* could have inflicted such an injury, and if it pierced arteries of the pelvic region, the moa would have bled to death. Using this technique, an eagle could swoop down and knock over a moa 10 times its own weight, then kill it by grasping and squeezing the animal's pelvis and neck, causing fatal bleeding.

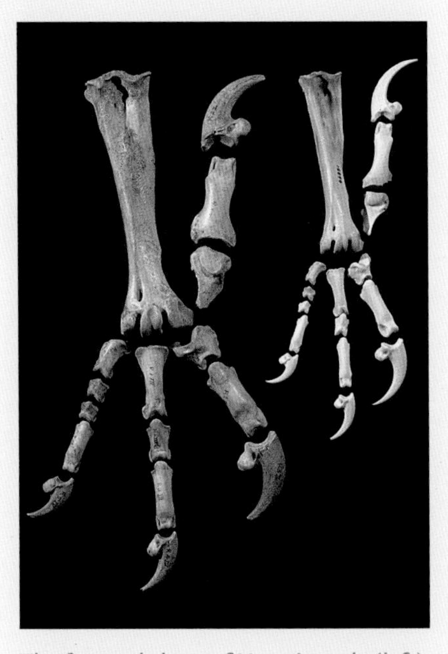

The feet and claws of Haast's eagle (left) compared with those of a wedge-tailed eagle (right), which is one of the largest eagles living today.

Other eagles are constrained in the size of their prey because they need to carry it away to eat in safety. With no mammal predators or scavengers to compete with, Haast's eagle had no need to carry prey. Like a lion at a kill, a bird could feast for days on large prey where it lay. This was virtually the only eagle in the world to occupy the very apex of the food chain. In other lands, top predators are the great cats.

That Haast's eagle became the largest eagle in the world is not so surprising when you consider that it lived among several species of moa, some of which were very tall and very heavy. And in any case, evolution on isolated island groups, such as New Zealand, regularly throws up giants. But DNA of the giant Haast's eagle reveals an unlikely secret about its ancestry.

In order to understand the evolutionary history of this fearsome eagle, in 2005 scientists extracted mitochondrial DNA from its bones and compared it with that of 16 living eagle species. Which of the suspects, they wondered, was its closest living relative? The odds favoured the wedge-tailed eagle of Australia and New Guinea, mainly because Haast's eagle was thought to have a large wedged tail. When the results came in, however, they revealed that the closest relative was the most unlikely candidate. It was *Hieraaetus*, one of the world's smallest eagles, whose closest living relative is the Australian little eagle. Furthermore, researchers propose that having been blown across the Tasman Sea from Australia on the westerly winds and establishing itself in New Zealand, the little eagle evolved into the largest eagle in the world in an astonishingly short time. Using molecular dating methods, experts estimate that it took little eagles just over a million years to evolve into giants. From the world's smallest eagle to the world's largest represents a tenfold increase in size! No other vertebrate in the world has achieved such a rapid rate of evolutionary change.

A female *Dinornis* moa was taller than a man, as seen in this museum display (above). The bones of this giant bird were found where it lay down and died, in a cave in the Honeycomb System in Karamea (right). On the mummified head and neck of an upland moa (left), even the tongue is preserved. Upland moa were the very last of all moa species thought to have gone extinct; this relic was found near Queenstown.

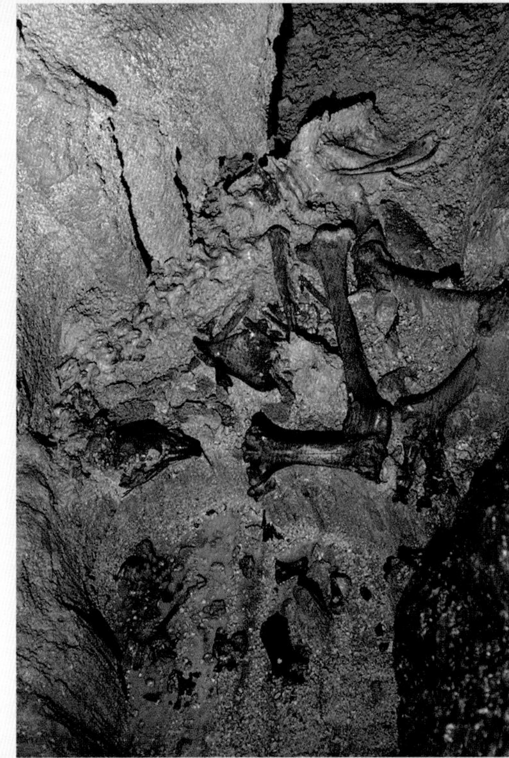

What stopped Haast's eagle growing even bigger was the physical limitations imposed by the demands of flight. At 15 kilograms, Haast's eagle had reached the upper weight limit of a flying bird. And though it had strong legs for moving around on the ground, it also must have used its wings to fly up to high perches from where it would swoop down on prey.

For many thousands of years, Haast's eagle ruled as New Zealand's official scariest creature. It lived almost exclusively in open forests in the south and east of the South Island. It preyed on the biggest moa species, but its main prey was most likely the large flocks of ducks that inhabited the uplands of the South Island.

So, why did the big bad bird disappear? Quite simply, another predator arrived in this land. Size, beak and talons were no match for stone weapons and fire. Cave paintings and myth tell of the mighty eagle known as te hokioi, which was named from the sound of its call. But it was not heard for long. Fires lit to flush out moa from the drier forests and scrubland on the east side of the South Island not only took away the giant eagles' prey, they also removed the habitat for both predator and prey. With eagle pairs requiring a territory that extended over hundreds of square kilometres, they could not have survived this double loss. The youngest eagle bones may be only a few hundred years old. In the nineteenth century, West Coast explorer Charlie Douglas described shooting something that could have been an eagle.

Was it the frightening presence of the eagle that is the source of the day terrors experienced by many birds? Is it a 'remembered' fear of this terrifying creature that still confines some New Zealand birds to being active at night? Kiwi, kakapo, even kea and kaka, are all at their most active beneath the protective cloak of night. We, on the other hand, prefer the protective cloak of the duvet, which is a major disadvantage as night-time is the best time to be out and about in our bush. Night is when our forests come alive.

This Maori rock drawing of Haast's eagle, from a cave at Pareora, near Timaru, is an almost life-size — if fanciful — representation of the great predator. Even more fanciful is the idea that this giant evolved rapidly from the smallest eagles in the world, yet it's true. Australia's little eagle (right) is thought to be the closest living relative of Haast's eagle.

Many native New Zealand creatures are nocturnal because they live in the shadow of the past. It is not only kiwi that avoid light. This one (right) emerges from her nest hole, which she will soon lay in. Kakapo (below) creeping around the forest floor after dark would once have been a common sight. While our reptiles and insects (far right) have no need to fear eagles, they do have a new nocturnal enemy — rats, which hunt by smell rather than sight.

Get out in the bush at night and, depending where you are, you are likely to meet several kinds of land bird; several species of burrowing seabird will be out and about, as will all kinds of weta, tuatara, native frogs, all the geckos, many of our skinks, and a full complement of all our native slugs, snails and flatworms.

If you can deny yourself the warmth of a sleeping bag for a few hours, a top tool for wildlife spotting at night is a torch attached to your binoculars. Your best option would be a system like the one devised by herpetologist/top reptile spotter Tony Whitaker. As 'Whit' says: 'Nocturnal geckos can be located by eye-shine when torchlight is reflected from the *tapetum lucidum* at the back of their eyes. Geckos' eyes are small and their pupils contract rapidly when light is shone on them, thus limiting reflections. Best results are obtained when the light source is as close as possible to the observer's line of sight, and when light levels are relatively low. This is best achieved by using a spotlight mounted on binoculars so as to give a coincident line of light and line of sight. What also helps is an electronic current regulator to enable light levels to be adjusted according to the range being searched.'

Sounds like the perfect birthday gift for a nocturnal wildlife spotter.

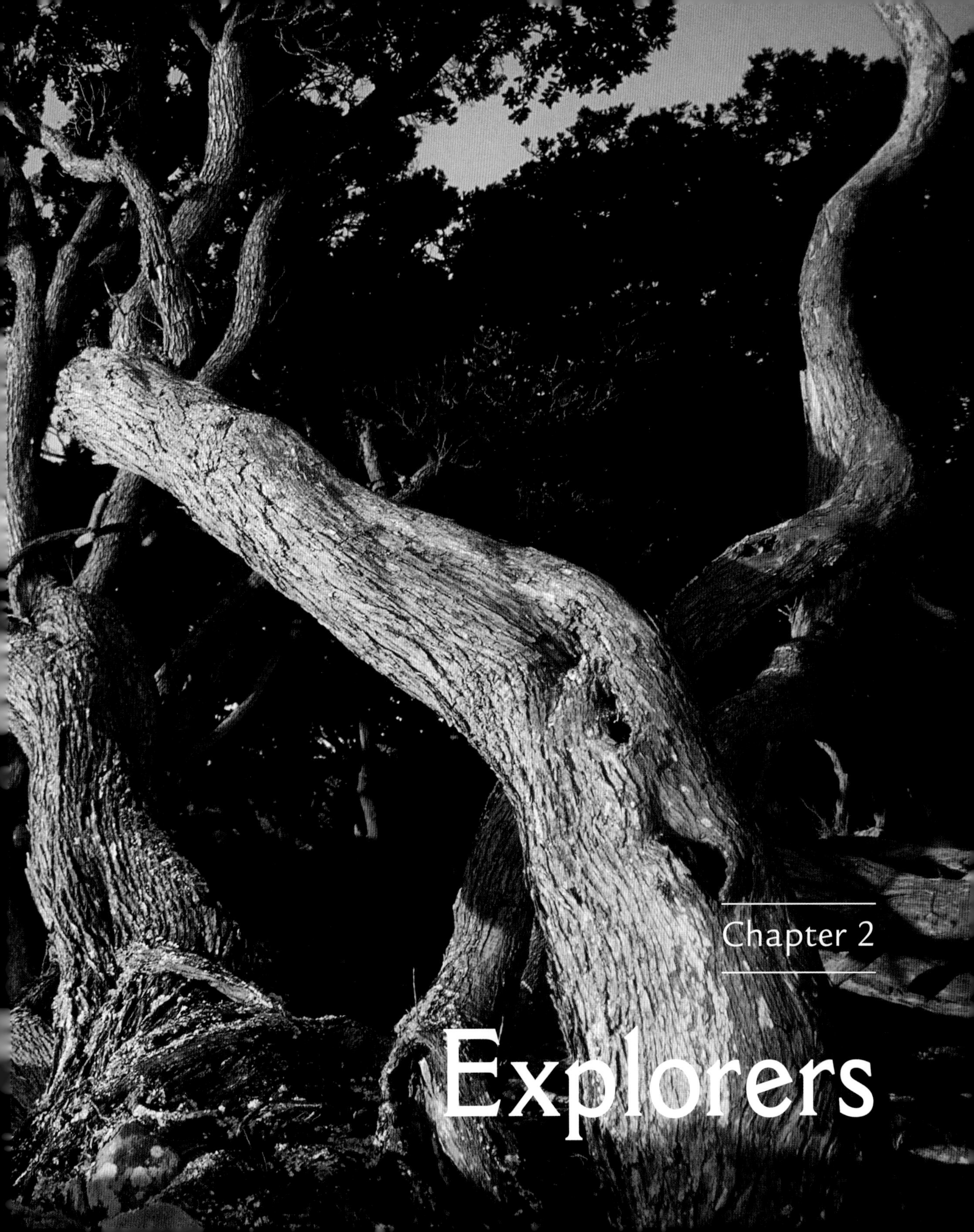

Chapter 2

Explorers

The town of Te Araroa, just a few kilometres from the North Island's East Cape, is a town with impact. On my first visit it wasn't impact as such but a near miss with a Cessna taxiing up the main street to have its tyres pumped up at the garage. The second impact was startling. Down by the beach, inside the fence of the Area School, I stood slack-jawed in front of Te Waha o Rerekohu, the largest pohutukawa tree in New Zealand. It was on fire with red flowers, 20 metres tall with imploring branches reaching out to 20 metres on all sides. Here the magnificent tree stood, a solitary giant, on permanent detention at school, looking out across the beach northwards to the tropical Pacific, where its ancestors have explored and colonised new lands like the navigators of old.

Question: who was the greatest explorer of the Pacific? The clock is ticking ... chances are you are thinking Captain Cook, or maybe Kupe. There are also votes coming in for Abel Tasman, Jules Dumont d'Urville, even some for Ferdinand Magellan, the man who gave the mighty ocean the name Pacific, the 'peaceful sea'. But another worthy contender should be pohutukawa. Yes, a plant as explorer. If pohutukawa had two legs and a passport, it would go down as one of the world's greatest explorers. Pohutukawa has spread out across the vastness of the Pacific, from the Philippines and New Guinea in the west to Hawaii and Tahiti in the east; from an island off Japan in the north to the subantarctic Auckland Islands. Cook and Kupe couldn't match such an epic achievement.

A major base for this exploration was New Zealand, which is home to several cousins of pohutukawa, all part of the genus *Metrosideros*, which means 'heartwood of iron'. The timber is beloved for its durability and strength by Polynesian and European builders of boats and bridges, flooring and furniture. It is burned as fuel and admired in art.

James Cook and his naturalists visited many islands of the Pacific where they identified *Metrosideros* species, relatives of our pohutukawa. On windswept summit ridges of Lord Howe Island (right) grows a *Metrosideros* called the mountain rose. At home, pohutukawa is often seen growing at the back of a beach, as here on Little Barrier Island (below right), or forming the entire forest canopy as on the Poor Knights Islands (below). Could *Metrosideros* from New Zealand be more widely travelled than the illustrious Captain Cook?

New Zealand's two most separated and celebrated cousins are pohutukawa and southern rata. Pohutukawa, *Metrosideros excelsa* ('the sublime *Metrosideros*'), is a northerner, its natural range beginning at Young Nick's Head near Gisborne, which lies some way south of the school yard at Te Araroa. From there it spreads north, hugging the coast and throwing a blanket over offshore islands, extending to the very top of the country, where the roots of Te Reinga, the tree at North Cape, guide spirits of the dead on their journey home to Hawaiki. On the west coast, pohutukawa forest remnants are still found down to a latitude north of New Plymouth.

At the other end of the country grows southern rata, an elegant and sculptural tree with stunningly beautiful flowers and true grit. Southern rata is virtually the only tree to survive on the bleak, storm-ravaged Auckland Islands, where it yields to the relentless pounding of storm winds by ducking down and leaning over almost horizontally. Where gnarly branches touch the earth they send down roots, and in sheltered glades yellow-eyed penguins and New Zealand sea lions nest and snooze.

Southern rata also claims seniority over its flashy northern cousin. Recent research points to southern rata being as old as the dinosaurs; a tree whose bold red flowers have blazed for almost as long as New Zealand's ancestral land mass has existed. It was way back then that this group of trees evolved to become great explorers.

The pohutukawa of the south is southern rata (below and right). Its forests cover the cool, wet West Coast flanks of the Southern Alps, but they also extend to islands in the subantarctic. Being able to withstand cold and wind makes them the last tree standing in the far south. On the Auckland Islands, southern rata forest provides a safe haven for yellow-eyed penguins, or hoiho (below left).

One place to see pohutukawa in action as an explorer/coloniser is a short ferry ride from downtown Auckland. Step ashore on Rangitoto Island and look one way and you'll see volcanic rocks as black and raw as if they'd erupted yesterday, not 600 years ago. Look the other way, and there's dense pohutukawa forest. A few paces up the track you'll meet the colonisers: small, solitary pohutukawa seedlings growing directly out of bare rock. Elsewhere, you'll see them sprout from bare cliff faces, sand dunes, cracks in paths or walls … even around Rotorua thermal vents!

Their roots are amazingly determined, curling and curving like miniature snakes hell-bent on finding water. Pohutukawa cousins survive like this in many tough, windswept, arid outcrops across the Pacific; and what's more, these trees are now known to have originated in New Zealand. But how did trees from New Zealand reach remote, isolated islands surrounded by vast tracts of ocean? They did it by using a family trick.

I live next door to a huge pohutukawa that stuns us with its spectacular flowering display in summer. Tui and bellbirds announce flowering as they squawk and shriek, claiming ownership of the tree. Flowers appear suddenly, then for two weeks it's all on, as birds and bees knock back 'all you can drink' nectar from the best bar in town. They pay for their drinks by pollinating the flowers. Just as quickly, flowering ends, the blooms drop innumerable red stamens and the tree begins the work of making seeds.

To explore the wide ocean, to find new lands in far-off places, the seeds must be travellers. I can vouch for that. Every autumn the pohutukawa next door sends its seeds over the fence. Tiny scratchy seeds cover everything; but this short journey from my neighbour's is nothing to a pohutukawa seed. A wind speed of 5 kilometres an hour or

Metrosideros has a number of tricks up its sleeve that make it a good coloniser. The seeds (above) are like dust and so light they can travel great distances on the wind. They are resistant to freezing temperatures and immersion in salt water. If seeds do find land, *Metrosideros* trees are one of the few that grow on bare rock (right). The hanging red beard (below) is made of aerial roots. As a tree ages and limbs droop, these roots penetrate the soil, extending the spread and size of an individual tree.

more is enough to keep seeds airborne to become playthings of the wind, and that's how they can spread out over the Pacific on truly epic adventures. DNA studies show that seeds from New Zealand are the source of forests in Tahiti, Fiji and the Solomon Islands. Our pohutukawa have even colonised Hawaii. That's an unfeasibly huge journey of over 6000 kilometres, which involves crossing the equator into the northern hemisphere.

New research conducted by Dr Shane Wright, a biogeographer from Auckland University, and a team of collectors led by Tony and Viv Whitaker, analysed samples of all the many species of *Metrosideros* on Pacific islands separated by huge expanses of ocean. In Tahiti and the Marquesas Islands, pohutukawa stake a claim to ridgetops, where the climate is more temperate and there is less competition from the tropical vegetation below, which is why the collectors found themselves clambering along eroded, knife-edge ridges, with sheer drops of hundreds of metres. And their perilous adventures paid off. Here they found some old friends far from their New Zealand home. Some of the trees looked a lot like their ancient cousin, the southern rata.

The four-year project identified several species around the Pacific related to New Zealand pohutukawa. What's more, the genes of some of them revealed that they'd made the great journey from New Zealand relatively recently — during the most recent ice ages, which ended a little over 10,000 years ago. This discovery led Shane and the team to develop a theory as to how those journeys might have happened. They studied ice-age climate models which showed how mid and low latitudes around New Zealand were buffeted by very powerful westerly winds. They theorised that these robust ice-age winds carried pohutukawa seeds to the Marquesas Islands in the remote eastern Pacific. From there, climate models could also explain the improbable journey of seeds north to Hawaii. During the ice ages, extreme southeasterly trade winds, powered by heat differential and the spin of the planet, could have drawn cool air towards the equator. So in two hops, pohutukawa seeds could have travelled from New Zealand, via French Polynesia, all the way to Hawaii.

Look one way on Hawaii's Mt Kīlauea and it's just like being

Steep, eroded, fog-shrouded ridges in French Polynesia, like those of Mt Marau, Tahiti, are home to the pua rata, the Tahitian *Metrosideros*.

During the most recent ice ages, New Zealand pohutukawa seeds spread across the equator to the islands of Hawaii. Here they dispersed widely, even colonising recent lava flows. So successful was pohutukawa in colonising Hawaii that several forest birds, like the crimson ʻapapane (below), a species of finch, evolved to feed on its nectar and pollinate its flowers.

on Rangitoto Island. There are pohutukawa trees and seedlings growing on black basalt. Look the other way, where molten lava flows in red rivers, and you realise it's not Rangitoto — Kīlauea is still active! Hawaii is the youngest of the Hawaiian Islands. On Maui and Oahu, 'pohutukawa' has been growing for hundreds of thousands of years and has evolved into several subspecies.

The tree from New Zealand has also found a place in the hearts and culture of Hawaiian people, who call the tree *ohiʻa* and the flower *lehua*, in honour of a legendary loving married couple. The story goes something like this: Ohiʻa's good looks and physique caught the eye of the goddess Pele, who passed his garden each day. She then decided to have him for herself. One day, while Pele was trying to woo the young man, his wife, Lehua, came out to give him lunch. Pele instantly saw the great love they shared and the futility of her attempt at seduction. Jealous and enraged, the goddess struck down Ohiʻa, turning him into a twisted ugly tree. Distressed, Lehua fell on her knees beside him. Other gods, looking down and seeing Lehua's heartbreak and despair, transformed her into a beautiful red flower so that the two lovers would never more be apart.

The extraordinary colonising prowess of pohutukawa and southern rata has not helped them survive back in New Zealand. Here, both have been battered, bruised and beaten in a sustained attack by possums. Possums can destroy even a mighty tree. They feed on its young shoots and leaves until it has no energy to regrow and it dies. Images of skeletal white 'ghost trees' on remote northern headlands and on the flanks of the Southern Alps are heartbreaking. Death on such a scale is possible because the possums are exploiting a family weakness.

Pohutukawa, southern rata and all their relations belong to a huge, ancient and powerful family called the myrtles. Myrtles are all around us. In our gardens we grow guavas and feijoas; in the forest there are several including manuka and kanuka. Australia is home to the best-known myrtles in the world, the eucalypts. All myrtles share family traits, such as spectacular flowers and an ability to tough it out in harsh conditions. Another myrtle family trait is having leaves that produce powerful aromatic oils, alcohols and other chemicals; the best known are eucalyptus and manuka oils with their therapeutic properties.

In Australia, ancestral home of eucalypts (and possums), eucalypts have in the course of evolutionary time made oils and compounds in their leaves that are toxic to possums and koalas. However, our myrtles did not evolve with mammal browsers, so the aromatic and chemical cocktail brewed up in the leaves of New Zealand's myrtles, particularly pohutukawa and rata, is largely toxin free — and possums adore it. Perhaps the taste is a reminder of home; like a piece of pavlova or an Anzac biscuit, with no nasty aftertaste.

Up in Northland many people still fondly remember the glorious pohutukawa forests that spread along so much of the coast. Memories dissolve into raw anger in recalling what happened so suddenly to change all that. Possums reached Northland in the 1970s and caused such mayhem and mass die-off of trees that by 1989, just 10 per cent of the original coastal forests remained.

The koala, which feeds only on eucalyptus foliage, can stomach the harsh toxins within the leaves. Marsupial and tree have evolved to tolerate each other. These spectacular pendant blooms of red-flowering gum (above) bring to mind the showy flowers of pohutukawa; and yet although *Metrosideros* is indeed their close cousin, it does not grow there. When possums (right) arrived here, two very different worlds clashed, and the encounter has been disastrous for pohutukawa. A similar disaster would befall the mountain rose (below) if possums ever got to Australia's Lord Howe Island (see page 29).

If you can find a pohutukawa or southern rata forest these days you will see another agent of the trees' destruction. The first thing you notice walking through a grove of trees is that it's easy going. The view is unobstructed because both species produce leaves that require high light intensity, and so are mostly crowded in the uppermost layer of the canopy. Flowers, too, are up aloft. The forest is easy to walk through, but having all your leaves on the outer margins makes trees particularly vulnerable to possum attack.

Possums love new growth, and a tree that has been done over by these marsupial menaces must put enormous energy into replacing all its young leaves and shoots. The trouble is, possums love regrowth even more. This is devastating for a tree, as it will keep trying to regrow; but after the possums have returned for the second or third time, the loss of photosynthesising leaves and buds deprives the tree of the energy to regrow, and death soon follows.

New Zealanders' love of the crimson blossoms of pohutukawa has led to a movement to safeguard them from possums. Project Crimson began by protecting trees that were becoming skeletons, like the grand old tree near Whangarei Heads (above right). Now, the movement has spread to the far south, where communities such as the pupils at Macandrew Bay School on Otago Peninsula plant southern rata trees.

Great efforts have been made to turn the destructive tide. In the South Island, some of the most magnificent southern rata forests are west of the Otira Tunnel on the Arthur's Pass road. In summer, whole hillsides glow neon red in afternoon light. The sight brings motorists to a stop and they gaze for long minutes at nature's magnificence. But this could easily have been hillsides of white skeletons, had a decision not been made 30 years ago to protect the Otira southern rata with 1080 poison. Regular aerial drops of this controversial poison are still happening, but without it we would not have an intact forest for kaka, tui, bellbirds, reptiles, bees — and us — to enjoy.

Up north, pohutukawa are deeply etched into the hearts of many people, and it was the loss of their shading arms along holiday beaches, and their softening profile on headlands, that stirred locals, iwi, professionals and a generous sponsor to get behind an organisation called Project Crimson. Their mission is simple: to raise money for possum control and to plant as many pohutukawa as they can on new sites and replace those that have been killed. The passion that Project Crimson members have for tree renewal is more than matching the possum's appetite for destruction.

Most Project Crimson planting has happened on the North Island east coast, but recently attention has swung west, where much pohutukawa forest was burnt or used for timber long before possums came along. Here, the rediscovery of another pohutukawa 'trick' now offers great scope for coastal conservation.

Between Kawhia and Aotea Harbour grow the last remnants of a vast pohutukawa sand dune forest. Only the tops of these multi-crowned old trees poke out of sand dunes; their trunks are buried many metres below. The ability of these semi-submerged trees to stabilise the ever-moving sand has inspired Project Crimson to replant local dunes with new seedlings. The next step is to put pohutukawa to work further along this expansive and windswept coast to do what it does best: face the sea, bend to the wind and go to work colonising a desolate shore.

Chapter 3

Cold Case

It may seem strange, but the basement freezer is a go-to place at the home of co-author Rod Morris. Children in particular are always keen to know what he might have lurking in its cold confines. They are wide-eyed with expectation as his wife, Erin, opens the lid and unwraps a newspaper parcel. It's like a scene in a crime thriller: who will the frozen victim be? A little blue penguin, perhaps a takahe, maybe a bittern or a saddleback? It's a chance to gaze at the stunning detail in their feathers, or their claws, eyes or beak. Each would have died by accident or from natural causes; and each will be immortalised as prey in a photograph featuring a falcon or some other predator.

Visitors are only persuaded to close the freezer by news that the refrigerator has live animals. A row of plastic boxes is home to several weta; a peripatus, the ancient velvet worm; and a richly patterned leaf-veined slug. Some creatures are so at home in the fridge that they have bred and raised babies there. Over many years of trying to keep invertebrates for his photography, Rod has discovered that most survive best in the fridge.

It seems odd that these creatures could be so comfortable and relaxed in a home appliance at a temperature that hovers close to zero. It only begins to make sense when we realise that these small animals naturally live in places where nights and winter temperatures drop to zero. And because so many of our plants and animals can withstand long periods of cold, scientists call them 'cold tolerant'. It seems they evolved this ability from ice-age ancestors. In the last two and a half million years, ancestors of weta, peripatus, native slugs and all the other plants and animals of this land have had to endure around 20 periods in the planetary fridge, many lasting tens of thousands of years.

All these animals have lived out their lives in Rod Morris's fridge. The bluff weta (above) enjoyed the freedom of free ranging, and the peripatus (right) lived there for many months. Residents also included leaf-veined slugs, whose young (below) hatched, emerged and grew up in the cool, moist environment.

It turns out that New Zealand's plants and animals had an extended training period preparing for the cold. It began around 12 million years ago when the two main tectonic plates beneath New Zealand began pushing against each other with considerable force. The land had nowhere to go but up, and the Southern Alps slowly juddered and shuddered upwards at a rate of up to 20 millimetres a year. Increased height was accompanied by growing pains in the form of spectacular erosion of rock and silt, which was swept away from the young mountains by broad and powerful braided rivers. But as the mountains rose, many plants and animals, which had existed in the tropical Oligocene prior to these upheavals, found themselves on a slow elevator heading skyward. And as the land rose to become mountains, it got much colder. They had to adapt or perish.

Barely had they adapted to life at altitude when, about one million years ago, the first of the Pleistocene ice ages began. (Scientists tell us we are still in the ice ages, but are currently in a brief interglacial warm period. Another ice age will come, they claim; the question is when? Some say in 50,000 years, others 10,000. Still others say it should have happened 2000 years ago but the agricultural revolution stalled it. Whenever it happens, our plants and animals will be ready.)

The last ice age ended a little under 20,000 years ago. Had you been here, you would have noticed big differences wherever you lived. You most certainly would have felt the cold, since the average temperature was four or five degrees cooler. But, the views would have been spectacular. In the South Island, vast ice sheets covered most of the Southern Alps and Fiordland and some seriously giant glaciers ground their way down river valleys out towards the coast. There were even glaciers in the North Island. During the most recent ice age, with much of the world's fresh water locked up in ice, sea levels were lower, and you could have walked

The Upper Hooker Valley in Aoraki/Mt Cook National Park. Much of New Zealand was like this during the ice ages, and much of our plant life, like the southern beech forest growing on the valley walls, thrived in these conditions.

all the way from the North Island to the South Island and on to Stewart Island. All that ice meant that plants and animals in cooler parts of this land had to evolve to endure freezing conditions, or, if they could, migrate north or towards the coast. Some may even have survived in slightly warmer sheltered pockets among the glaciers.

The last ice age began releasing its grip on New Zealand around 20,000 years ago. Ice-core records show that 13,000 years ago, levels of heat-trapping carbon dioxide in the atmosphere began to rise, and our glaciers began a rapid retreat. Scientists estimate that glaciers lost more than half of their length and mass in just 1000 years.

A thousand years is a blip in geological or evolutionary time. With the retreat of the ice, and a rise in temperature, there was an abundance of new land and opportunities, and our plants and animals didn't have to disperse far to be part of this warm new world. Plant pollen and beetle skeletons reveal that during the coldest part of the ice age, small patches of forest survived in most parts of frozen New Zealand. And when the ice retreated, life advanced.

Some species were too specialised to move. Alpine weta stayed behind, as they had evolved to survive the worst of the cold in a remarkable way. On a snow bank on the flanks of the Southern Alps, spring sun melts the ice. At first it reveals a rock, then beside it a mountain weta, which has spent the winter entombed in ice. But as the sun plays on its apparently lifeless body, the frozen body cavity thaws; within and between cells, ice turns to water. Finally, when there is sufficient warmth, one antenna twitches, then the other. Then a miracle happens — the weta begins to move. Within a few minutes it is feeding on a hebe, its first meal since being smothered in snow four months ago. How did it survive?

Ice kills. As a body freezes, ice crystal shards grow within cells, tearing at their walls and organelles. However, some hard-core plants and animals have evolved to survive freezing. They possess special 'antifreeze' chemicals that prevent destructive ice crystals forming, so that when thawed, undamaged cells can restart their life processes. It is as if all winter their bodies have been on pause. The technical name for this state is torpor.

A massive glacier in South Georgia in the South Atlantic (above). During the ice ages, New Zealand's own glaciers (below with kea) covered much of the land, extending out to the coast. Today, in the mountains, we get a glimpse of how ice-age fauna coped. Scree weta (right) emerge from winter dormancy to feed on hebe shrubs even before the snow has retreated. Over 3100 glaciers larger than 1 hectare still exist in New Zealand as reminders of the big chill.

The recently awoken weta scuttles away just in time to avoid becoming breakfast for a kea, which instead nibbles on the seed heads of a nearby hebe shrub. It isn't fussy; kea eat a huge variety of food types, which is a necessity for survival in the mountains. Meal over, the kea takes off, its Ferrari-red underwings glowing in the morning sky as it swoops down-valley.

So why did a parrot become a mountain specialist when it could have escaped the cold by flying north? This bird is an extraordinary enigma. We call it the mountain parrot, but there's evidence that its range used to extend almost to the sea. Perhaps we have forced it to the upper edge of this former range. Many high-country farmers dislike kea. More than dislike, they hate the birds because they kill lambs and weakened adult sheep. In earlier days, they were shot on sight. There was even a bounty on their heads. Back in 1885, when two shillings a beak was paid in the Mackenzie Country, 1400 birds were shot in one year. Little wonder they have retreated back up into the alps to live the life of mountain parrots. Perhaps if kea hadn't been persecuted, we might have called them grassland parrots.

In early summer, below the permanent snowfields, New Zealand's flower garden comes into bloom. Hillside after hillside is covered in buttercups, eyebrights, orchids and daisies. There is no wind, no sound, except for the repeated clackety percussions of cicadas. These cicadas are dark coloured, with wings shorter than those found in the lowlands. There are several species in this *Maoricicada* genus and together they are the most alpine cicadas in the world. It is thought they were once lowland cicadas, but that over the last two to three million years mountain uplift has put them on a slow elevator ride up into the alpine zone, where they survive by sucking plant juices.

There are many animals and plants in New Zealand that, like *Maoricicada*, can cope with the cold; but research shows that there is one animal that can't — a penguin. We tend to think of penguins as creatures of snow and ice. How can a penguin be averse to cold?

The rise of the Southern Alps, and the ice ages that followed, led to the evolution of high-altitude specialists like mountain daisies (above), alpine cicadas (below) and, of course, the kea (right). To survive in a harsh environment the kea became resourceful. Its strong, hooked beak is a vital tool to dig for starchy roots of alpine plants. Unfortunately, this tool, coupled with high intelligence, has led it to kill high-country sheep (below far right) and create conflict with farmers.

Kayaking with Hector's dolphins in Akaroa Harbour can be enormously frustrating. You catch a peek of the curved dorsal fins and paddle towards them, only to lose sight of them, until they reappear far across the other side of the bay. Soon, you give up and hang out with a little blue penguin that is lolling about on the surface, casually preening the waterproof feathers that give it the buoyancy of water wings. These mini penguins are a delight. A scratch of the head reveals a white flipper — this is its ID. The penguins of Banks Peninsula are called white-flippered penguins and recent genetic research shows that they are closely related to the blue penguins further north in New Zealand and those on the Chatham Islands. However, it seems that Banks Peninsula is a blue penguin frontier and that blue penguins further south, in Otago and Southland, are quite different animals.

The reason for the difference could be closely linked to little blue penguins' dislike of cold. It's thought that the blue penguin frontier became established during ice ages, when the ice was thickest and glaciers were longest. Back then, according to research, little penguins disappeared from anywhere south of Banks Peninsula; it was simply too cold for them.

The smallest dolphin (above right) and the smallest penguin (below right) in the world are both found around the coasts of New Zealand, at locations including Akaroa Harbour, and are unique to this country. While both have relatives that cope well with cold, these 'mini' dolphins and penguins are intolerant of it. White-flippered penguins (below) are distinguished from little blue penguins by the greater amount of white on the leading and trailing edges of their flippers.

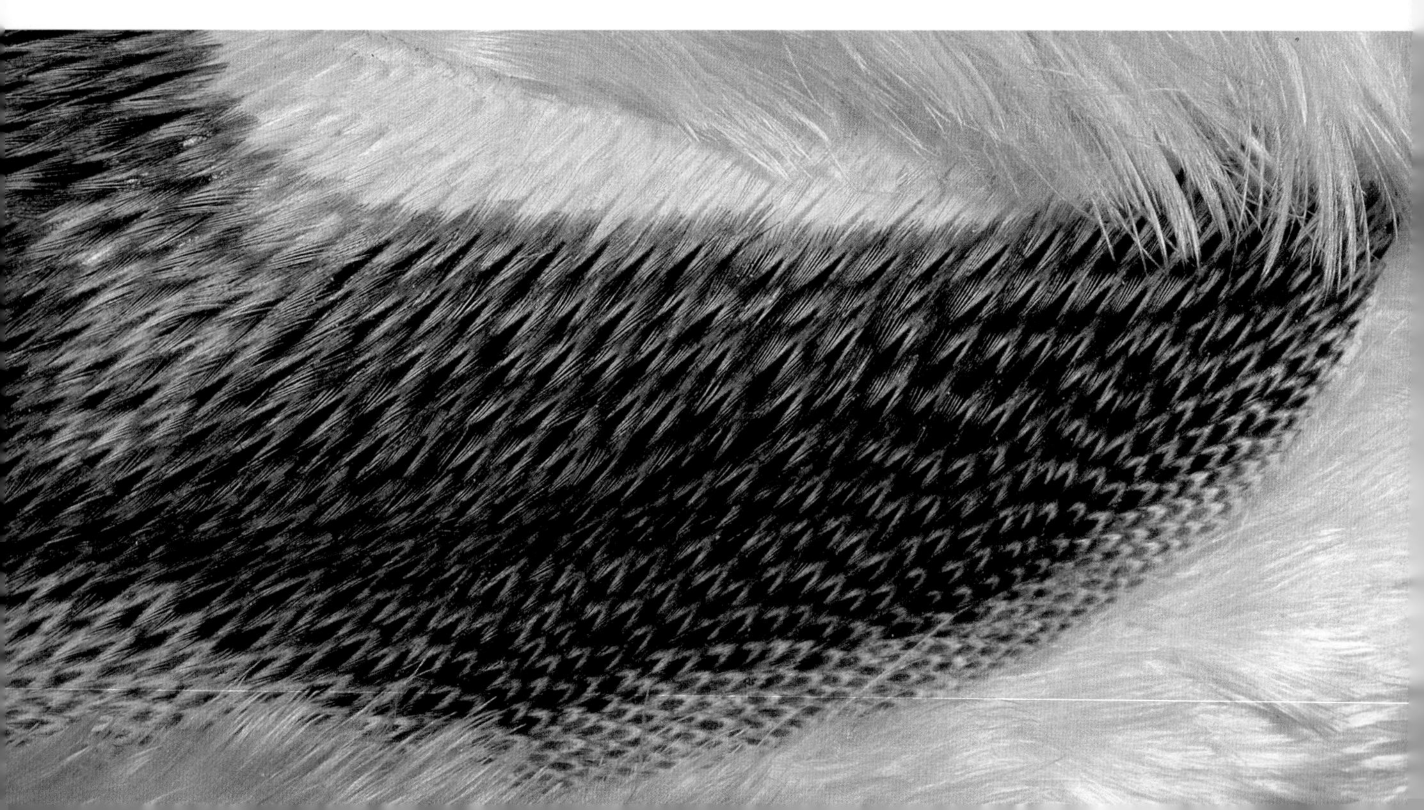

Recent genetic data suggests that the southern form of the little blue penguin (these pages) is in fact the Australian fairy penguin, which possibly established itself on our southern coasts after the ice ages. The southern birds are a brighter blue than the little blue penguins that extend from Banks Peninsula north. In a good breeding season they produce up to three clutches of eggs, enabling them to increase numbers rapidly.

But when the ice receded and the weather warmed, it wasn't the white-flippered penguins that spread down from the north to colonise Otago, Southland, Stewart Island and around to Fiordland. Instead, according to recent genetic research, blue penguins from Otago and Southland carry genes that are more closely linked to Australian blue penguins. Could Aussie blues have colonised our southern shores?

According to one theory, in the thaw after the last ice age, when it was once again warm enough for blue penguins to occupy the southernmost parts of New Zealand, the birds that moved in were those that swam all the way from Australia. It is not an impossible journey; westerly winds and favourable ocean currents would have helped them on their way.

The idea of Aussie blues inhabiting southern New Zealand is supported by the fact that the blue penguins down here look more like Australian blues and their calls seem to have a hint of an Australian accent. It's remarkable that they have kept their identity even though they probably crossed the 'ditch' to colonise our southern shores as long as 100,000 years ago.

However, there is another, more recent, theory emerging that states that it might have been more recent human hunting pressure, not ice, that helped drive the little penguins from southern New Zealand to extinction, thus allowing the Australian blues to gain a foothold in the south.

Which theory is correct? As any detective will tell you, sometimes you can re-examine a cold case using more modern methods — but the trouble is, which modern method do you go with? Perhaps time will tell.

If the hunting pressure theory is proved correct, it could possibly rewrite the story of the demise of other native species, such as the takahe and Eyles' harrier; and the subsequent arrival and successful colonising of other recent Aussie immigrants like the pukeko and the swamp harrier.

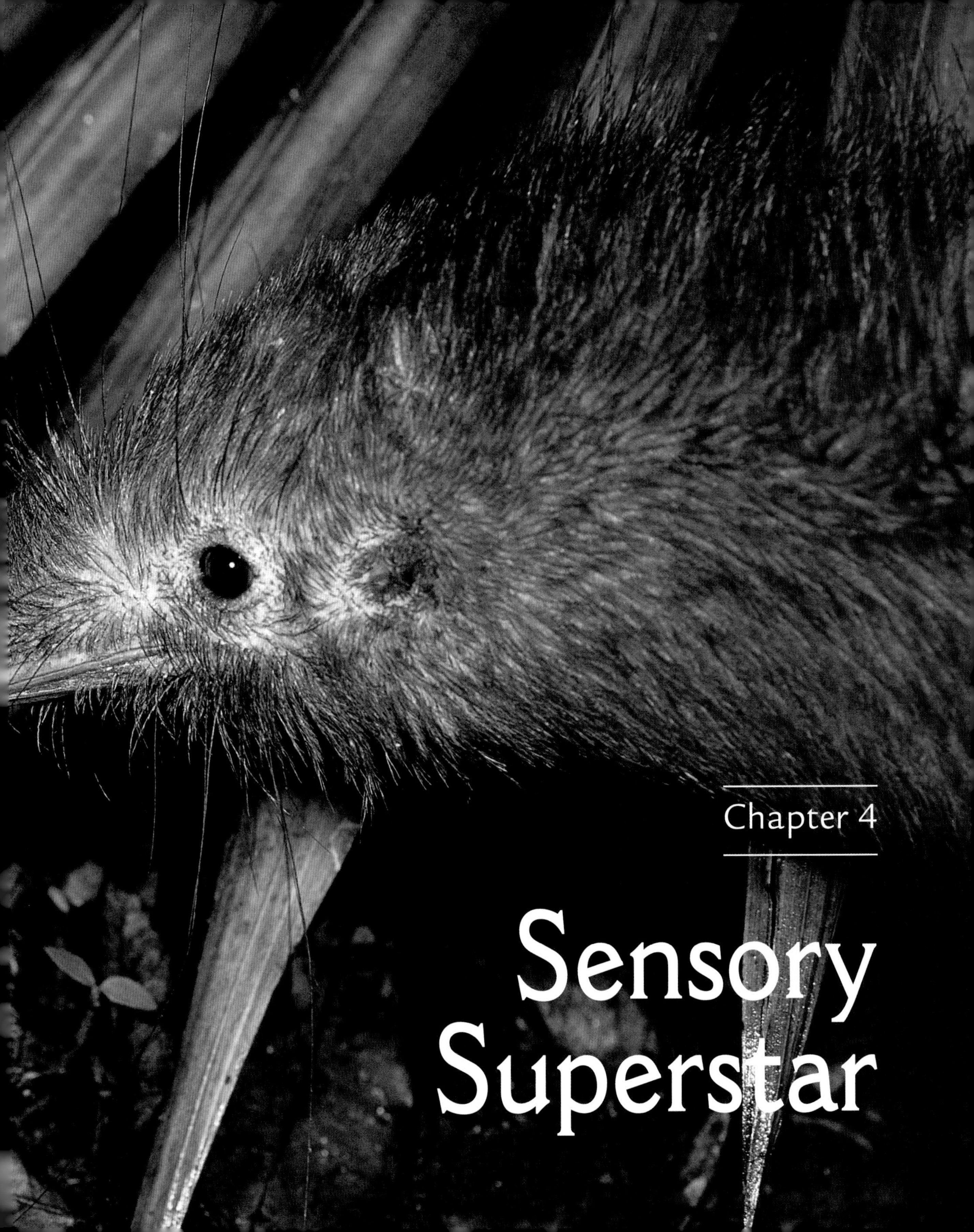

Chapter 4

Sensory Superstar

It was one of those summer days when tar sticks to your jandals. The ocean and the pool on Napier's Marine Parade were so inviting, but a swim was not an option. We were on a mission to the new Kiwi House. Inside was as cool and dark as midnight. Daytime eyes slowly made sense of dim details — ferns, fallen logs, a clay bank, shrubs, and beyond, something moving, a hunched shape with a long beak. It moved a little, feeding, I guess, but was too far away for me to be sure.

This was my first 'face to face' with our national bird, and I was unimpressed. Why wouldn't it come out of the shadows? We waited and waited, shivering in the cold. I craved warmth and sunlight. The kiwi seemed unreal, unfortunate and unknowable. How wrong I was, about that and everything else I took away from my first kiwi encounter.

I have met kiwi many times since and have come to love them for their eccentricity. No other bird in the world comes close to this package of evolutionary surprises. The kiwi is unique in having an extremely low metabolic rate, offspring that hatch with adult plumage, and nostrils at the tip of its beak. Its 'oddness' separates the kiwi from 10,000 other bird species in the world.

Kiwi are also unusual because they are experts at a job for which there was great opportunity in New Zealand. Kiwi were the most skilled members of a group that raked, dug, probed or gleaned invertebrates from the forest floor. When New Zealand was one vast forest — that is, before we arrived — the forest floor was the place to be.

The kiwi is a strange bird, with a shaggy coat that provides perfect camouflage (above). Its long, probing beak (below) enables it to feed on soil-dwelling invertebrates that in other countries are normally hunted by burrowing mammals such as moles, shrews and badgers. Right: A young Haast tokoeka emerges from its subterranean home looking like a burrowing mammal.

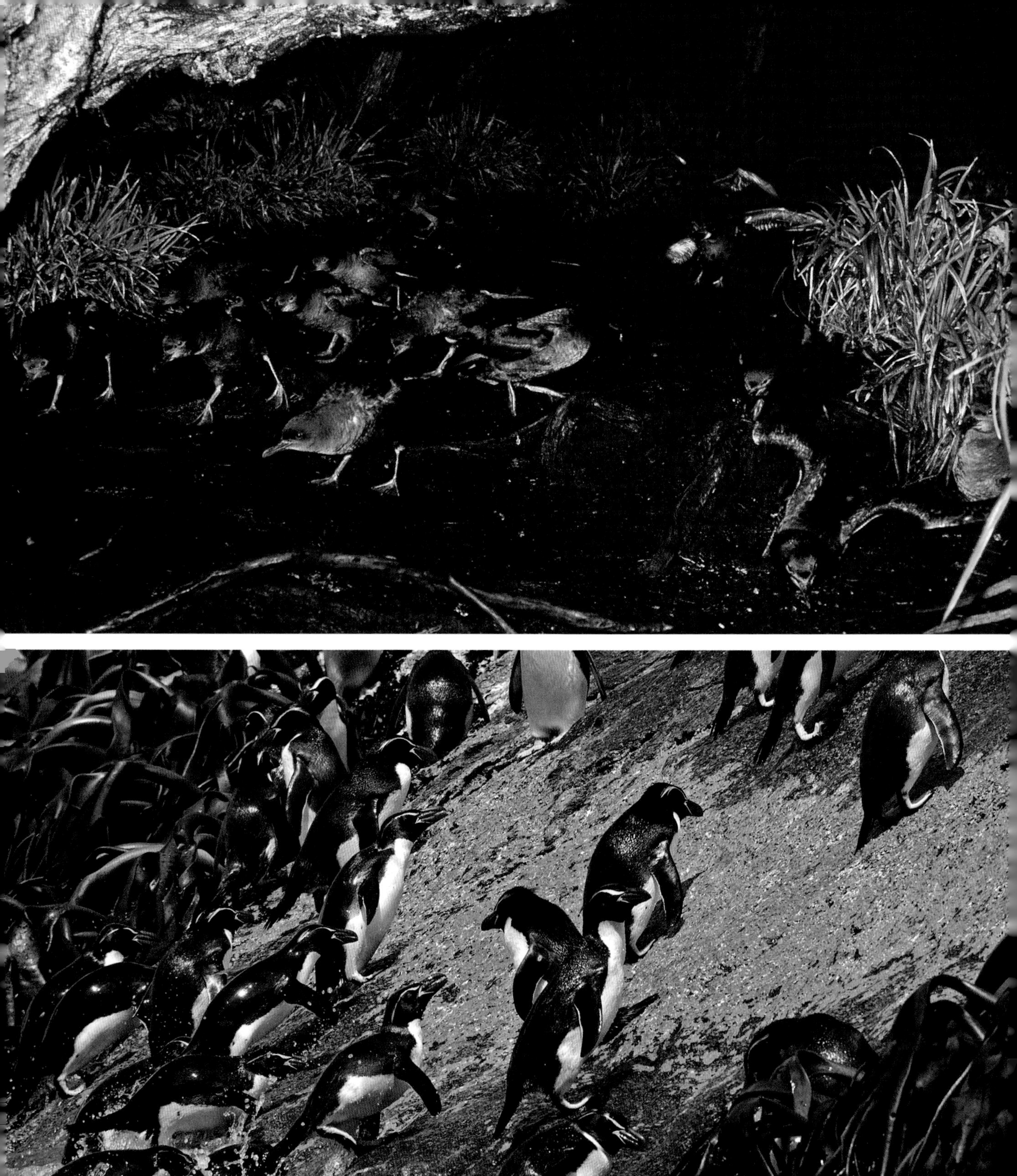

On the Snares Islands, south of Stewart Island, we get a glimpse of what much of ancient New Zealand must have been like. At night, millions of shearwaters crowd the forest floor (above left) and fill the night skies (above); by day, Snares crested penguins come and go (below left). Both shearwaters and penguins bring the riches of the sea back to the island to feed their chicks. They also bring tonnes of fertiliser, which enriches the forest and soil, resulting in an abundance of invertebrates.

The only place we can get to imagine the glory days for forest-floor insectivores is on predator-free, forested islands such as the Snares Islands, south of Stewart Island. At dusk, millions of sooty shearwaters return from fishing, circling the forested island against a blood-red sky. Each evening, they bring two special gifts from the sea to the island. One is partly digested fish, which they feed to their ravenous chicks; the other is their droppings, rich in nitrogen and phosphorus from their seafood diet, which fertilise the mineral-hungry soil. A daily dose of tonnes of fertiliser feeds plants, which in turn feed many small creatures such as beetles, flies, worms, snails, spiders, ants, weta, woodlice and cockroaches that live in the leaves, soil and streams.

There are no kiwi on the Snares Islands to snaffle up all the bugs in the soil and leaf litter, but there are tutukiwi, or snipe. A snipe is the size of a blackbird and has a long, kiwi-style bill. It probes busily for insects around shearwater burrows, like a kiwi on speed. Snipe are related to wading birds called sandpipers. They may be small, but they have pulling power, and the reward for successful probing may be juicy worms and other creatures, which they pull from the damp soil. Snipe are now found only on predator-free islands, where they create a scene from the past that would have been very familiar throughout New Zealand.

In old New Zealand, a large 'guild' of birds and animals benefited from the prolific invertebrate life in our soils. Many of these creatures are small and rare. The snipe (above left) was once found all over the main islands, but today survives only on predator-free offshore islands. The ground-foraging short-tailed bat (below left) is confined to old-growth forest. Others like the owlet nightjar (above; this one from Tasmania) and the bush wren (below) are extinct.

The Snares Islands, just 3.5 square kilometres in area, are pock-marked with three million sooty shearwater burrows and serves as a nesting site for several other seabirds — imagine the same density of seabirds on every coastal plain, headland and hillside in all of New Zealand! This is how it was before humans arrived; billions of shearwaters, gulls, terns, petrels, penguins, cormorants and albatross in vast colonies. And every day, those billions of birds would have fertilised the land, making it invertebrate rich and highly desirable for the insectivore group.

Unfortunately, many ground-dwelling, invertebrate-eating birds became targets of the mammal predators that were introduced by settlers. Never again will we see the snipe rail, a little weka-like bird with a long beak; the owlet nightjar, with a huge opening mouth; or the laughing owl, a large predator that ate large invertebrates and competed with kiwi. Surviving invertebrate-eating birds include snipe, robins, wrens and saddlebacks. Non-avian examples include short-tailed bats, tuatara, skinks and geckos. However, the kiwi is probably the most specialised invertebrate eater of the lot.

This specialisation helps explain why kiwi are so weird. Look around the world at other animals that hunt invertebrates on the ground at night. You'd most likely come up with a short-list of mammals like hedgehogs, tenrecs, shrews, moles and badgers. The kiwi is one of the few birds dedicated to this kind of work. It is a superbly adapted creature that we are only now beginning to appreciate and understand.

A lot of new information has come from Massey University scientist Isabel Castro and her students, who study brown kiwi on Ponui Island in the Hauraki Gulf. They estimate that brown kiwi spend around 75 per cent of their night hours foraging for invertebrates, and while away the rest doing things that we might do — commuting, property maintenance, personal hygiene, raising a family, and so on.

It is not easy to observe kiwi in the wild. The research team get around that problem by using a night-vision video camera. Their crystal-clear infrared images reveal that what kiwi get up to in the dark is often strange, and even remarkable and quite unexpected. At night they emerge from their burrows to feed. They search for invertebrates in leaf litter, rotten logs and even under water. They use their long beak for probing underground for worms and the larvae of many insects. One native worm glows brightly as it's pulled from its hole; another, over a metre long and as thick as your thumb, involves a real tug of war, but provides a huge one-course meal.

Until recently we thought kiwi detected prey by smell, but then scientists discovered a honeycomb of cells at the tip of the bill, used to detect vibrations made by the movement of prey. These combine with the bird's senses of smell and hearing to help it navigate in the dark. We, on the other hand, are near-blind at night; without sight, our other senses are of little help in detecting even large objects in our path (above left).

The kiwi's other nocturnal activities are more difficult to observe and understand. One of Isabel's students, Susan Cunningham, suggests that to really get to know an animal that lives in a world without light, you need to 'put yourself in their situation'. That's the key to understanding. So I decided to give it a try: to put myself in a kiwi's shoes, as it were. That meant heading outside at midnight, blindfolded. The scientists reported that kiwi use familiar tracks to move around their territories, so I decided to use my own territory, the garden, in my attempt to be a kiwi. But before I set off, I waited. Why? Because we are such a sight-centred species, we relegate our other senses to the second tier. Without the use of my eyes, I needed to wait for my other senses to step up. First to kick in was hearing; a complex soundscape of close, distant and far-off sounds flooded my senses. Then came a heightened sense of smell; I hadn't realised what a smelly world I inhabited. I was becoming a kiwi! To be fair, though, although it was amazing to tune in to these underutilised (or masked) senses, after just 10 minutes in my garden I could hardly claim to have superpowers.

Kiwi have one of the largest brains for body size of any bird.

It's up there with the smartest parrots. Recent research points to kiwi being able to use their brains to integrate information, cognition and learning. We use a lot of our brainpower to see, but kiwi seem to use their brainpower for processing smell, touch, vibration and sound from sense organs grouped around the face. Their senses of smell and hearing are extremely powerful.

Okay, so the kiwi had an advantage; but as I set off on a circuit of my territory, I found that my other senses were integrating, working together as a team. Hearing road traffic on one side and the sea on the other provided instant direction, but I was also tuning in to smells that helped pinpoint different parts of the garden. I hadn't realised that flowers put out so much scent at night, as do compost, soil, the sea ... even concrete has a smell. There were so many aromas. Next came touch: the brush of plants, the surface beneath my feet, these helped keep me on track. Ouch! Damn wheelbarrow left on the path.

Experiments (not ours) with kiwi have shown that when an obstacle is placed on regular trails, a bird may bump into it a couple of times, but thereafter its behaviour changes. Next time, it will stop before the obstacle and quickly detour around it, before resuming its journey along the path.

I was now lost. Smells were ever changing and confusing; a kiwi has smell beacons along the track, while all I had was an informational arm-wrestle. By now I was on my hands and knees, calling on touch and taste to help me. The garlic patch was a good beacon, but further on I got flummoxed; was I tasting kale or hydrangea? Was I heading for the back door or the fence? I was totally lost in my own garden.

I had tried to explore the dark world of the kiwi, and failed. But then, the kiwi can also call upon another sense, a secret weapon that makes it a super-sleuth in the dark. Recently Isabel and her team discovered that not only can kiwi smell using their nostrils, but in the honeycomb of bone at the end of the bill is a cluster of cells, called Herbst corpuscles, that detect vibrations produced by prey moving underground. The cell cluster is thought to come into play when the kiwi taps its bill on the ground, and that when the bird picks up a vibration it continues tapping and feeling in a small area, eventually zeroing in on its exact location. Only then does it plunge its bill into the ground and extract the prey.

Kiwi (right) live in a world of sound and smell. Hearing paints a picture of their wider night-time world. Their response to unusual sounds is to freeze as their brain processes the information, and to remain still until danger has passed. This response works well in an ancient forest but can be fatal in a rapidly changing world of dogs, cars and other modern dangers (above).

The kiwi may be a superb secret agent of the night, but that has not helped it in the battle for survival in recent years. All five of our kiwi species are in decline; one, the rowi, or Okarito brown kiwi, we very nearly let slip through our fingers. It was renowned kiwi scientist John McLennan who first realised not only that the species was down to critically low numbers, but also that the survivors, all 300 of them, were 'old buggers' — some with bald patches, or one eye, a limp, a bad cough. It was a kiwi old folks' home. More frightening was the fact that there were no young birds to be seen. What had happened to them?

McLennan and his fellow researchers discovered that chicks were hatching from eggs, but were being summarily killed by stoats. The lack of juveniles was a New Zealand-wide problem. Ninety-five per cent of all juvenile kiwi on the mainland fail to reach adulthood, but rowi chicks were exceeding the national average by 5 per cent, which meant that no rowi juveniles at all were surviving to become adults. Something had to be done — and fast.

One vital bit of knowledge was the discovery that once young kiwi reach 20 weeks of age and around 1200 grams in weight, they can defend themselves against stoats. That's how Operation Nest Egg (ONE) came about. Eggs and young chicks of North Island brown kiwi and Okarito rowi are collected and taken to a safe place to be raised to subadult size before being released when big enough to fend off predators. And it works. ONE is now used to restore failing populations, establish new ones, and maintain numbers in years when predator control fails to offer adequate protection.

There's little doubt that the kiwi would have long ago gone extinct like so many other birds, such as moa and huia, were it not for one thing: its nature. Kiwi are tough, gritty and bold. When cornered they can kick and scratch their way out of trouble. However, their stroppy nature cannot rescue them from a major decline in numbers. There were an estimated 12 million kiwi before humans showed up; now there are just 70,000 or so. While this may sound a lot, it is the total for all five species, and some of these, particularly the rowi, are in very low

A ranger (below) holds our rarest kiwi, an old rowi from Okarito. When it was found that their small population were all old and there was no recruitment of chicks (below right), Operation Nest Egg stepped in, removing eggs or young chicks from dangerous situations and rearing them in safety until they are big enough to defend themselves against predators such as the stoat (above far right).

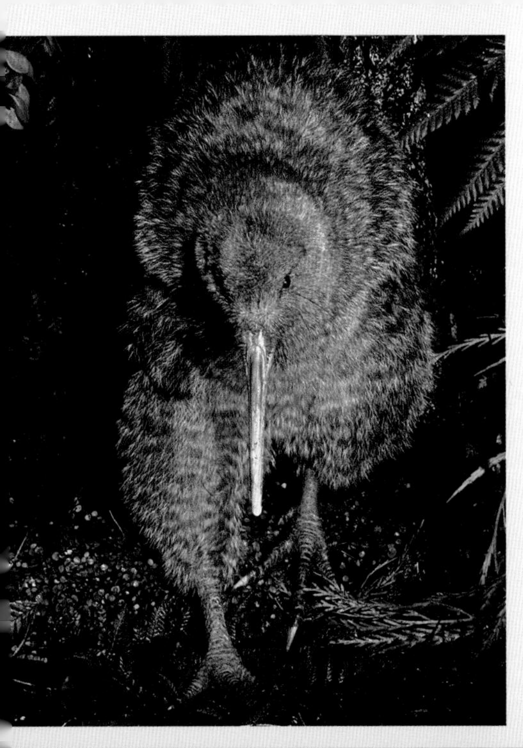

Great spotted kiwi (left) are the toughest and most aggressive of all kiwi species. They live high in the mountains, and adults are capable of looking after themselves. A young great spotted kiwi, however, is as vulnerable to predators as any other chick.

Opposite

Little spotted kiwi are our smallest kiwi. They are very vulnerable to predation and were the first to become extinct on the mainland. Fortunately, once placed in a predator-free sanctuary, they 'breed like rabbits'.

numbers. Experts agree that without help, and at the present rate of decline, rowi too will be gone in 75 years.

In the battle for kiwi survival, these tough little birds have one other trick at their disposal — but, again, they need our help to pull it off. If we can clear an area of predators, kiwi will breed prolifically. A dozen little spotted kiwi were housed at Zealandia, the large predator-free sanctuary in Wellington. They began to breed like rabbits. In this establishment phase, females were laying two eggs a year. The same thing is happening for other kiwi species in other predator-free sanctuaries, mainland islands and similar protected places up and down the country.

ONE is holding the line in some places and among some species, but there is another initiative to help kiwi that is growing rapidly in parts of New Zealand. Ordinary New Zealanders, farmers, businesses and local governments are coming together and getting behind kiwi conservation in a hands-on way. It's happening because we now have the knowledge to understand how they got into so much trouble and how to help them get out of it. No country in the world embraces its national bird as New Zealand does, to the point where we have taken its name. And now the kiwi bird needs a helping hand from us Kiwis.

Throughout New Zealand there are a number of community-based breeding programmes. A keeper at Willowbank Reserve (below) holds a young brown kiwi chick being raised in safety, thanks to Operation Nest Egg. At Arthur's Pass (below right), transmitters are replaced on a pair of great spotted kiwi as part of a community-based predator control and kiwi breeding programme.

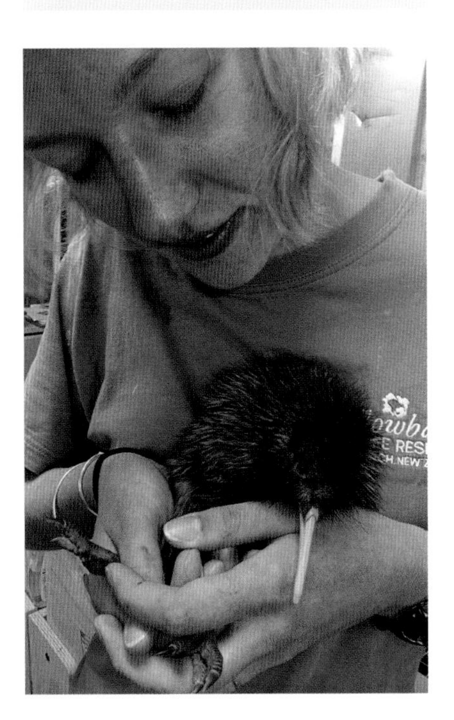

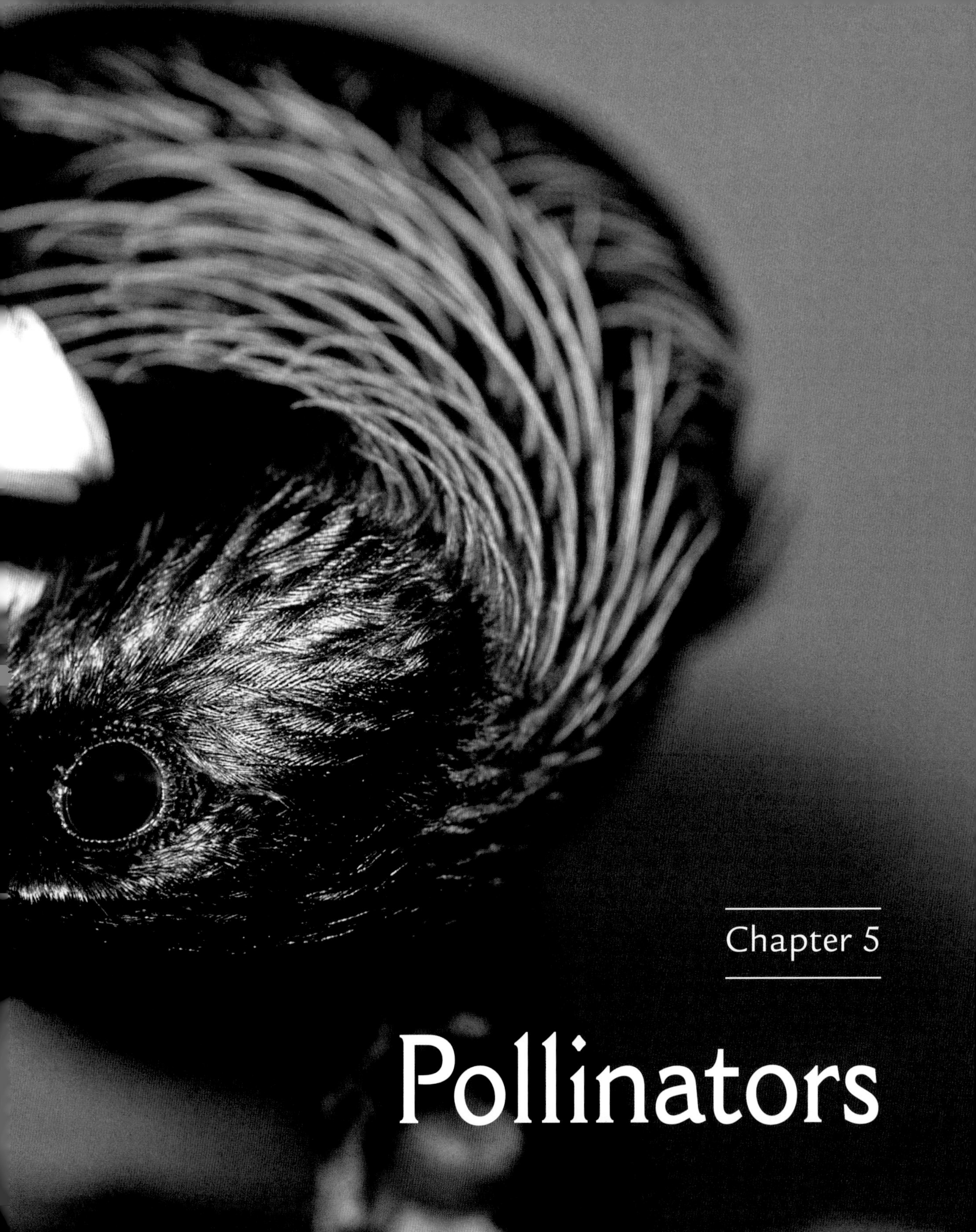

Chapter 5

Pollinators

In Invercargill in summer, if you are lucky enough to watch a tui feeding on a flax bush, pause and appreciate it. Not just for the delight of seeing the bird's forehead dusted bright orange with flax pollen each time it pushes its curved bill and brush-tipped tongue deep into a flower, but because, chances are, that tui has flown all the way from Stewart Island for the nectar reward it receives. The reward must be worth expending all that energy flying across the 20 kilometres of windswept Foveaux Strait, or the bird wouldn't do it. It must be worth it for the flax plants to run a confectionery factory during the flowering season, or they wouldn't do it. And that's the point. The deal that plants struck with animals to pollinate their flowers, signed nearly 100 million years ago, is as vital and important to both signatories as it has always been.

The first ever flowers were pollinated by wind. While red beech flowers might be small and easily missed (above), it is difficult to overlook the pollen over a beech forest on a windy day (above right). Later, plants formed an association with insect pollinators. Among the earliest pollinating insects would have been beetles (below right).

The pollination partnership must be one of the sweetest evolutionary deals between plants and animals. In return for unlimited access to the plant's dessert trolley (nectar), the animal helps the plant to fertilise other plants, seeing they are rooted to the ground and can't travel to do this. In New Zealand there are some nifty, even bizarre, ways of clinching the sweet deal on which many plants and animals have come to rely.

Plants do have other agents, such as wind, to help move their pollen around. Hay fever sufferers know only too well that come spring, wind-blown pollen from grasses and pine trees induces a chorus of sneezing and the checking of use-by dates on inhalers. Plants that use the wind to pollinate must mass-produce pollen in the hope that some random grain will come in contact with a random stigma. While this is a useful mechanism in windy places where there are plenty of plants of the same species, in other circumstances wind is an almost useless way to spread pollen. Evolution is a great problem solver, so a more efficient mechanism in less windy situations would receive an evolutionary tick.

So it came to be (note the vagueness here, because it may have occurred a number of times) that plants evolved devices that produced much less pollen, along with mechanisms to transport it to where it's needed. One of the most successful devices is, of course, a nectar-laced flower, and flowers now come in all manner of shapes, sizes, colours, textures, and

with many kinds of aroma. The first carriers of pollen were probably beetles, but as flower power grew, other creatures, including bees, wasps, moths and butterflies and even some reptiles and mammals, all took advantage of the sugar-for-pollen-transport deal. Every group in the animal kingdom, it seemed, was in on the action. Today, nectar and/or pollen is the only food consumed by many hundreds of insects, and 20 per cent of all insects at some stage of their lives depend on nectar.

The next step was exclusivity, as certain individual animal and flower species co-evolved the means whereby only one pollinating animal could receive the nectar and pollen of one species of plant. In New Zealand, this exclusive arrangement can be elaborate, requiring knowledge of a secret code.

It is always a pleasure to spot a mistletoe blazing red amid the sombre green of a South Island beech forest. (That pleasure is becoming rarer due to the possums that hungrily chomp these tasty plants, which grow on the branches of the beeches.) The mistletoe is a hemiparasite, which means it inserts its roots into a host beech tree's branch or trunk and freeloads on the tree's root system for water and nutrients. However, the mistletoe does have leaves and therefore makes its own sugars, and the entire arrangement seems not to harm its host. In summer, during the flowering season, mistletoe welcomes members of an exclusive club of birds which know the code that will open the blooms and give access to the nectar within.

A beech forest may be a purely native zone, but in summer the trees are visited by a cosmopolitan procession of native and introduced birds and insects. Many, including blackbirds, sparrows, starlings, honey bees and German wasps, are drawn to try to open the mistletoe flower, but since they lack the key to its treasure, the bloom remains resolutely shut. Only when a bellbird or a tui comes along can the secret be revealed. Just as only King Arthur could free the sword Excalibur from the rock, thereby gaining the right to become King of England, so only native honey-eaters (tui and bellbirds) have the means to drink deeply of mistletoe nectar and be anointed with its pollen.

Red mistletoe is a hemiparasite: it makes some food by photosynthesis, but obtains the rest from its host. Its flowers can only be opened by an exclusive set of pollinators. Tui know how to 'tweak' the unopened flower, causing it to spring open, giving access to the nectar reward within.

Another member of the special 'sect' of mistletoe pollinators is this tiny, solitary native bee (*Hylaeus*, left), which has a unique method of entry to the hidden treasure: it makes a small slit near the top of one petal, causing the flower to spring open.

A bigger native bee (*Leioproctus*, above) does not know the secret code and is confined to visiting flowers that have already opened and likely to have been raided by birds.

The visit and the opening of the flower happens quickly. The code is deceptively simple; all that's required is to give the tip of the flower a tweak or a twist and the petals spring open, like a twist-top bottle. Then, as the bird plunges into the nectary to claim its sweet reward, pollen-laden anthers catapult forward, spraying it with pollen. Such a brilliant design. The bellbird gets an untapped source of nectar, and the mistletoe protects its precious pollen until feathered 'Arthur' comes along. But there is another unlikely Arthur. A tiny native bee has unlocked the key to the mistletoe's secret stash.

A female native bee gives new meaning to the word worker. Unlike introduced honey bees, she spends her short adult life as a solo mum constantly on the move. After mating, she builds a nest in the ground, in which she begins to construct cells. As each cell is completed, she provisions it with nectar and pollen before laying a single egg in it. This is where the mistletoe joins the story. Females of some bee species have their own trick to access mistletoe nectar and pollen.

A visiting bee makes a small slit near the tip of the flower (about one in 10 of these cuts is in the right place) and magically pops one petal partly open, allowing it to squeeze inside. It then sets to work gathering all the pollen off the four anthers. As it goes about this work, it sometimes brushes against the stigma, dropping pollen as it does so, which allows the seed to ripen.

It's hard work, but this food helps the bee mum to provision a cell, lay an egg and move on to constructing the next cell. Phew! After six to eight weeks all her busy work is done and she dies, probably of exhaustion. But thanks to her hard work, and with some help from the mistletoe, she leaves behind 10 to 20 fat larvae, now developing into overwintering pre-pupae. Next spring, a new generation of supermums will hit the mistletoe flowers.

If we are to believe stories from over half a century ago, forests of today are blessed with far fewer flashes of red mistletoe to delight our summer senses. Possums, unless heavily controlled, remain mistletoe killers. But the scarcity also reflects how rats and stoats are causing a sad and dramatic decline in the number of honeyeater birds. Because only they know the tweak-and-twist secret of opening the remarkable mistletoe flower, most blooms fall unopened and unpollinated, and fail to produce seed.

Even riflemen (right) know the access code to open mistletoe flowers. However, this ancient, intimate partnership between the flower and a select few birds and bees is under threat from introduced predators, such as the stoat (above). In some areas, stoats have greatly reduced the number of birds that can pollinate mistletoe. This in turn has led to most flowers failing to be pollinated and falling to the ground unopened.

The pollination story gets even stranger and darker in the night forests of the central North Island. It concerns another plant parasite and a distant relative of the vampire bat. The plant is called the wood rose, or pua te reinga, 'the flower of the underworld'. Intrigued? I was more than that: I was beside myself with curiosity when I was first taken to see this flower in Pureora Forest near Rotorua.

The wood rose is so named because it parasitises the roots of forest trees, and where it 'plugs in' to the tree it forms a gall or burl on the forest floor that

looks somewhat like a wooden rose. Unlike the mistletoe, this flower of the underworld receives all its nutrients from its host. The wood rose has no green leaves; it is a total parasite.

My first impression of the dull-coloured flower clumps poking out just above moss level was less than enthusiastic. But that was soon to change. I was with a film crew setting up to shoot the visitors that would, we hoped, be pollinating these flower clumps after dark. Before we left for the long night vigil, I got down on my knees and had a sniff … yes, I got it, a faint musky smell, a bit like the funk in a men's changing room. It was definitely 'blokey' and lined up with what I'd read about this strange flower; the exact same alluring chemical is used by tamarin monkeys in tropical South America, and by ring-tailed lemurs in Madagascar to mark their territory. This seemed quite odd.

When I finally viewed the footage from that night, my impression of this plant instantly changed to total fascination. The images were black and white, recorded with infrared light, light beyond the range of all eyes except the camera's. Out there was as black as a bat in a hat.

The forest floor was still as the time-code numbers on screen sped on to 10 p.m. This was when the first movement appeared. Slowed to real time, the image showed a tiny bat approaching the flower. With wings folded up, it used its forewing as a pair of front legs and marched along on the forest floor towards the flower from the underworld, which was in the foreground. Other bats soon marched up. This was what the flower's boy-smell scent was attracting: not a monkey, but our only land mammals, bats. The short-tailed bat is very special; not only does it prefer to walk than fly, but some scientists think this bat and this flower have evolved a special pollination partnership, one that is just as exclusive as that shared by the mistletoe and honeyeaters. It's a good deal for the bat, because the flower of Hades is plugged into the roots of the tree, which means it makes copious amounts of nectar.

Wood rose is a root parasite with its own suite of unlikely pollinators. Filming the flowers at night revealed that this weird flower growing straight out of the soil had a range of pollinating visitors including tree weta and short-tailed bats.

During its brief flowering season, a pohutukawa is smothered with bright red, open brush flowers. This sends a powerful message to nectar lovers from near and far. Pollinators attending the nectar orgy range from kaka and tui (left) to weta (below) and even bats.

Take a midsummer stroll when New Zealand's Christmas trees are in flower and you will be perfectly placed to witness another pollination partnership in full swing. The cloak of red that covers pohutukawa trees, and the urgent chimes from the throat of a bellbird proclaiming ownership rights, show that the tree is open for business. The bellbird song stops when a tui arrives. All manner of other birds fly in, but sparrows, starlings and goldfinches lack special brush tongues for slurping nectar, and they struggle to get a drink on the wind-tossed clusters of flowers. The tui shows how it's done. Its brush tongue easily laps nectar from the little cups that surround each ovary, and each head movement also picks up a few more pollen grains from the sacs on top of the anther.

Silhouetted above the tree, a scribble of bees runs random relays from tree to tree, hive to flower. Bees, too, have a good technique for slurping nectar. There's plenty for all, thousands of vivid red blooms, each bloom made up of a single soft red female style, a column tipped with a pink knob where pollen is deposited. At its base is a ring of 20 to 30 male stamens, each with a double sac of pollen on top. Thus the flower is hermaphrodite. Between style and stamens is a shallow moat constantly filling with sweet nectar. Out on the end of branchlets, flowers are bunched so tightly that birds virtually walk over a carpet of stamens, and to get a drink, they thrust their heads down into the pile of that carpet.

On islands of the Hauraki Gulf, the overwhelming redness of pohutukawa sends a powerful message to us back on the mainland. The message reads, this is what I can do, this is how many flowers I can produce if there are no possums around.

Further north, the Poor Knights is one of these summer-red island groups famed for its marine life, which gathers in spectacular quantity where tropical and temperate ocean currents meet. Less well known, but equally spectacular, is its forest life. Beneath the mantle of these trees, tuatara hunt giant centipedes, which hunt large skinks, which hunt wetas. It is a world of mini giants.

Back in the 1960s, on a summer trip to the Poor Knights, herpetologist Tony Whitaker discovered that pohutukawa out there have an unusual night-time pollinator: geckos.

Even reptiles take on the pollinating role of the birds and the bees. As a Kaikoura gecko (above left) feeds on mountain flax nectar, it regularly licks sticky flax pollen from its face. A Poor Knights gecko (below left) feeds on pohutukawa nectar, and (below) a gold-striped gecko feeds on nectar from nikau flowers. A flowering bonanza attracts not only pollinators but also predators, like this giant centipede (above).

Tony was fascinated to observe not one but two species of gecko emerge at dusk, usually from retreats on the ground. On some nights, they were on the flowers before dark. He thought this might be because of competition for nectar. They seemed to gather around newly opened flowers, presumably because these have the greatest nectar production. As many as five geckos could be seen feeding together from a single flower. Their busiest nectar feeding time was the few hours just after dusk. When feeding, the geckos climb over the flowers and push their heads down between the stamens to lap up the nectar. On some trees there were as many as eight geckos per square metre.

On this one trip, Tony was able to open our eyes to a new pollination partnership. But recently, a researcher has extended our knowledge and understanding even further, and revealed the importance of yet another night-time pollinator. One night on Little Barrier Island, just after dusk and the din of petrels returning to their burrows, PhD student David Pattermore became aware of much activity above his head in an ancient pohutukawa tree. It was like a party going on up there. Not wanting to flash his torch and scare away the party-goers, he put on a pair of night-vision goggles. The goggles amplified moonlight to reveal an eerie black-and-white view of hundreds of short-tailed bats coming and going from the flowers.

From that encounter, David's study became focused on using infrared light to film bats feeding not just on pohutukawa but a variety of other tree species. What he discovered is nothing less than astonishing. The downy coats of bats are perfect pollen collectors, and their technique of crawling over flowers and plunging their heads in to get nectar spreads the pollen very efficiently. But it was their energy and speed that most surprised him. Feeding time on each flower was barely a couple of seconds, after which they would leave, not for a nearby flower but a nearby tree. Within a minute, the bats were back on the original flower. Such speed and mobility makes short-tailed bats the perfect night-time pollinator. Who knew?

In other experiments, David guarded some flower clusters from the many birds on Little Barrier that pollinate pohutukawa by day. He also excluded bats from other clusters. The difference in the amount of seed that was set — in other words, the success of pollination — was dramatic. Excluding birds made little or no difference to the overall amount of seed produced. Excluding bats resulted in a dramatic decline. His research shows only too clearly that having lost bats from most of New Zealand, we have also lost a major pollinator of many forest plants. It also forces us to reconsider the role of bats in the country's overall ecology. We talk of New Zealand being a land of birds, with our native bats as a quirky aside. Perhaps, thanks to David's research, we should change our thinking and call New Zealand a land of birds and bats.

Native short-tailed bats enjoy the most varied of diets. They are equally at home on the ground eating earthworms and dragonflies (right), as high up in the trees eating fruit. And, as researchers have recently discovered, they can play a vital part in the pollination of many New Zealand trees and shrubs.

Chapter 6

Bob

Down in a stream bed that trickles through a city suburb, young explorers are on the lookout for creatures as ancient as dinosaurs.

In a quiet corner of a pool, two feelers begin to sway; two beady eyes appear, revealing a well-camouflaged 'crawly', or koura. As the small crayfish walks slowly forward on four pairs of legs, its world is suddenly turned upside down. It uses reverse thrust to escape, but to no avail. In a jar, it just goes round and round backwards. A large distorted eye comes close. The female koura has babies clinging to the underside of her tail. They will remain there until big enough for independence. The young explorer gently tips this mother of many back into the stream.

In a laboratory, another koura is under a microscope. Gently, tweezers remove a tiny white creature from the joint in its front claws. It is a flatworm called *Temnocephala*, which lives on the koura but does it no harm. This microscopic creature is one of several that play exceptionally large roles in defining and explaining New Zealand's geological history.

What makes the koura and its hitchhiking flatworm so special is that they cannot tolerate salt water. This means that their ancestors could not have swum here across a salty sea. Also, they have close relatives in Australia, New Guinea and South America. These two facts are a powerful argument — some would say living proof — that all of these widely separated lands were once joined together. Scientists call this method of getting around by staying put 'vicariance', which is the opposite of the process called 'dispersal', whereby animals swim, fly or float to a new land.

It seems that some of our freshwater fish species could have done both. Some are vicariant (stay put and let the land move), others are dispersers (get out and swim to a new land). That discovery, along with many others, was made by a towering figure in the study and understanding of our native freshwater fish. His name was Bob McDowall.

Koura are often the keenly sought-after quarry of children out fishing. However, eminent fish scientist Dr Bob McDowall discovered that the koura, and its hitchhiking flatworm, plays an important role in explaining our geological history. Neither can tolerate salt water, so they were here long before New Zealand became a group of islands.

Bob, who died in 2011, said that the many discoveries he made during his scientific career were marked by a series of what he called 'aha' moments. The first 'aha' came while studying the redfin bully in Wellington's Makara Stream. One day, he saw them spawning. Bob noted that the males guarding the nests were a velvety black colour in contrast to their more colourful red-finned look during the rest of the year. He remembered that, when fishing as a boy while on holiday in Taupo, he was never able to catch black common bullies — and now, as he studied male redfins, he discovered why. Males not only turn black during breeding, but they also stop feeding. 'Aha' moments such as this led Bob to become a world expert in a group of freshwater fish. Through them he discovered a great deal about New Zealand's biogeography (the study of why species live where they do, and how they got there) and the story of the birth of these remarkable islands.

The male redfin bully is one of the most brightly coloured of our freshwater native fish. During the breeding season, however, it transforms into a velvety black fish that does not eat and so is very difficult to catch.

Introduced trout, like this male rainbow trout, are popular with anglers, but their diet includes high numbers of our smaller native freshwater fish (below). Bob McDowall was one of the first to make this discovery.

Bob's first job out of university was with the Marine Department. His laboratory was in the old Wellington city morgue, not an auspicious start. His supervisors directed him to study the diet of trout, which he reluctantly agreed to, but in reality he worked under the radar on native freshwater fish, for which there was precious little information available. On field trips to the Waikanae River, he began to work out the ecology of an extraordinary group of New Zealand fishes, the galaxiids, and their young, which we know as whitebait.

Bob had a passion for discovery and spreading the word about our unique freshwater fish, but there was one argument he never won, even with the weight of science behind him. That argument lies behind a deceptively simple question: what are whitebait?

Through the years, many people — from scientists and politicians to public bar know-alls and, of course, whitebaiters — have advanced theories to explain where whitebait come from. Some claimed that whitebait were the young of sea fish such as hapuku, kahawai, mullet, herring, minnows, pilchards, anchovies or sprats. Others suggested they were maggots.

Many New Zealanders are passionate about whitebaiting. But for many years we had no idea what was in our patties. By growing whitebait into adult fish in aquaria, Bob McDowall discovered that they could be any one of five different species of galaxiid: (from top to bottom) banded kokopu, short-jawed kokopu, inanga (see also above far left), koaro and (see page 135) giant kokopu.

By the 1960s scientists suspected that whitebait were the free-swimming and shoaling larvae of fish that lived in fresh water, but proof was elusive. It was Bob who led the search for the answer. He found it by studying samples of whitebait caught in the runs that took place in 1963–64. His team found that some were white and others were golden; some were thinner and others squat; some had forked tails while others did not. He kept live samples in aquaria and watched how they developed. This gave him the definitive answer.

All the fish that developed from whitebait grew into one of five different species of native river fish. They are collectively known as galaxiids, because many of them have tiny spots on their upper surface that reminded an early observer of the stars in the galaxy. The most common of the five species is the inanga, an elegant fish with skin, rather than scales, and a torpedo-shaped body. The second most common is the koaro, also called the climbing galaxiid. Koaro use large, ridged pectoral and pelvic fins like alternate suckers to clamp onto wet rocks of fast-flowing streams as they make their way up into streams in the high alps. Bob's samples also produced smaller numbers of banded kokopu, whose beautiful gold and brown banding provides perfect camouflage in forest streams. The smallest and rarest of the five is the short-jawed kokopu. Largest is the giant kokopu (in fact the largest galaxiid in the world), which inhabits slow, peaty lowland rivers and lakes.

Question answered? Well, not quite. A few diehard white-baiters still don't believe Bob's answer, trusting instead in their own instincts and ideas. The arguments can still be heard in pubs on the West Coast.

But the whitebait answer led to more mysteries for Bob to unravel. One of these seemed simple at first but turned out like a Russian doll, with one answer leading on to another and another. It began with this question: why do whitebait go to sea? All five whitebait species mature in different reaches of rivers and streams, and then in autumn males and females come together to spawn. Inanga differ in that they gather in tidal wetlands close to the sea. I have even seen them gathering in vast numbers in damp paddocks, under a full moon. Each female may lay thousands of eggs on grasses or rushes, which males duly fertilise, the water turning white with milt. When they hatch, the tiny larvae swim to the surface, where the outgoing spring tide sweeps them out to sea. The larvae of the other galaxiids also journey to the ocean. The question is, why?

For some freshwater river fish, sea water can be a toxic environment. Salt draws water out of every cell until the fish becomes extremely dehydrated and dies. Other fish, including New Zealand's galaxiids, are far more tolerant of the change from fresh to salt. They can prevent dehydration because they are able to control water loss. However, their cells must work hard to maintain a salt level that is suitable for life but is much lower than the salt level in the ocean around them. All this effort must be worth it or they wouldn't do it. So, what's the gain for whitebait galaxiids heading out to sea for the winter?

Spending part of one's life in a river and part of it in the sea is known as diadromy. Salmon are perhaps the most famous exemplars of this behaviour. Of the 18 species of galaxiid in New Zealand, half of them split their lives between river and ocean. But why do they do it? Again, Bob's research comes to the rescue. The answer is complex, but basically it comes down to food. Galaxiids spawn in autumn. At this time there is very little plankton in the rivers for their young to eat, but the sea is a vast food hall, with ocean currents moving plankton along the coast like a Japanese 'pick your dish' conveyor-belt restaurant. If they can survive the tough life at sea, larvae do well here. Then having fed and grown into their whitebait form in the ocean during the winter months, they begin the journey back upstream in spring. What is the signal to return? Not even Bob could supply a definitive answer to this one. But he suspected, based on whitebait catch records on the West Coast, that one trigger to return to the rivers could be snow-melt floods, which push huge volumes of fresh water out to sea.

Unlike whitebait, several galaxiid species do not go to sea. Among these 'stay at home' fish are the pencil galaxiids. They are small, very slender species found in cold gravel rivers. From top to bottom: bignose galaxiid, dwarf galaxiid, upland longjaw, alpine galaxiid. Many of the pencil galaxiids are widespread across central and southern New Zealand, with the greatest diversity in the inland Mackenzie Basin (right). Their distribution was influenced by ice-age glaciation and more recently by the taking of water for hydroelectricity and irrigation.

The habitat of the lowland longjaw is restricted to small rivers of North Otago. DOC officers Daniel Jack (foreground) and Ciaran Campbell carry out a monthly water quality check to ensure the survival of our most endangered native freshwater fish. Their most serious threats are poor water quality and predation by trout.

Bob was also intensely interested in galaxiid species that don't go to sea. Among these stay-at-homes there are some strange and unusual species, none more so than the so-called pencil galaxiids. Members of this little-known group are all small, some barely bigger than a matchstick. They are also very thin, a necessary shape for making a living by picking small aquatic insects from under stones in pools and in the gravels of stream beds.

His research indicated that four of the five pencil galaxiid species lived in the Mackenzie Country of inland Canterbury. Bob's interest in biogeography led him to theorise on the reason why most pencil galaxiids are found in one area of the South Island, and it hangs upon a broad understanding of the 'Oligocene drowning'. Around 23 million years ago, in the Oligocene period, Zealandia (the ancient continental mass of which New Zealand is merely the emergent part) was at its lowest height above sea level. In fact, some scientists theorise that the whole of Zealandia sank. But Bob's theory points to the tiny pencil galaxiids as evidence of species that could not have survived at sea, so there had to be some land and rivers — specifically, the Mackenzie Country — above sea level during this time.

Pencil galaxiids are missing from large areas of New Zealand, due, it is thought, to the effects of mountain building, vulcanism and mighty glaciers that elbowed their way from the Southern Alps out to the coast during recent ice ages. The lowland longjaw pencil galaxiid is the rarest, and may well have been pushed out by glaciers to the coastal rivers of North Otago. Part of the reason for its rarity is because its home rivers are used for irrigation, which in summer reduces water levels and raises water temperature, thus depleting the water of oxygen. But the little lowland longjaw has a trick that has perhaps helped it avoid extinction. When water levels go down, so do the fish. Recent research shows that they can burrow deeply into river gravels, surviving perhaps metres below the river, in underground flows that are a significant part of bouldery east coast rivers.

Finding how this fish survived could well have provided Bob McDowall with another 'aha' moment.

Chapter 7

Curious

How do you discover a new species? It's an outrageous idea. Surely we know what's out there? Actually, we don't, not by a long way. The latest estimates of total species on this planet — that's everything from the blue whale down to the tiniest one-celled plankton species that the blue whale sucks into its mouth — are somewhere between five and 30 million species (and about eight million of them are insects). Of these, we have named and described two million, so there's no shortage of opportunity for would-be species hunters.

The traditional view of a species collector is the Victorian naturalist explorer with a funny hat returning from far-off places with copious boxes and jars of exotic creatures. These days, most new species are discovered by experts armed with high-tech equipment, decent funding and a good idea of what they are looking for. However, there's another less well-recognised group that has made some stunning discoveries. I am referring to children. Their only qualifications for making new discoveries are age and innate curiosity.

Today, children are more aware of the global threats to the environment – but, for many, their physical contact with nature is fading. This has been called 'nature deficit disorder'. Getting outdoors brings other benefits. It encourages journeys of curiosity and imagination – in the mind, as I discovered as Davy Crockett (lower right); in the wilderness (upper right); or in the garden, as Jamie Morris discovered (below).

So, why didn't I become a child naturalist explorer? I think I was in the wrong place. My earliest wild places were high in adventure, but low on new species. We would spend days in an overgrown vacant section, making tunnels in the grass and wild fennel. Later, we moved on to the creek, catching eels and climbing pine trees. We encountered lots of blackbird nests and boxthorn, but nothing new to science. To succeed, I needed broader horizons, real wilderness. Let me illustrate what a young explorer can discover out there in the wilderness, armed only with a good dose of curiosity. This tale was told to me by co-author Rod Morris.

Rod has two sons. Jamie is now in his thirties, but when he was eight, he and younger brother Sam went with the family on holiday to Fiordland. They camped in a valley surrounded by sheer-sided mountains, which echoed with the crash of falling rocks in summer, and in winter, the roar of avalanches. Not your normal holiday camping spot. But Rod was on a photographic assignment and his subjects were black alpine insects. There are many hairy 'all blacks' in our alpine zone. Their dark colouring helps soak up the sun's energy more efficiently in a climate zone where temperatures can suddenly plummet. Rod was in luck: in just a few days, he had got some great shots of the black mountain ringlet butterfly and its caterpillar. He had also shot some alpine grasshoppers, cicadas and a spectacular speargrass weevil.

Jamie and Sam weren't short of things to do. They played with kea that gather near the entrance to the Homer Tunnel, which provides the road link to Milford Sound. They also fossicked among rocks nearby. That's where Jamie first spotted a very weird-looking fly. It was black, with wings like Batman's cape. He rushed back to tell his dad, who told him that he was busy and would take a look later. It turned out to be a lot later, when they were packing up to go home, that Rod found time to take a look.

A few minutes up the track Jamie pointed to a largish black insect basking on a rock. Rod had never seen anything like it, and he certainly couldn't identify it. It was a fly but it had the wings and size of a small butterfly. He photographed it, caught it and took it back to Dunedin, where local entomologist Tony Harris identified it as *Exsul singularis*: the bat-winged fly.

Like the caped crusader himself, this weird creature was a mystery. It had not been seen for many years, and so little was known about it that it was regarded as the rarest fly in the world. As word spread, Jamie was surprised and delighted to read coverage of his discovery in most of New Zealand's daily newspapers. It even appeared in *New Scientist* magazine. His major disappointment was that he'd been misreported. They wrote he was five years old, and he definitely told the reporter that he was eight!

Rod Morris would often pay his two sons, Jamie and Sam, pocket money to find bugs to photograph. On holidays in the Southern Alps they found alpine cicadas (above); short-horned grasshoppers (above right) and speargrass weevils (below right). All exhibit dark coloration that is a feature of alpine insects, a lesson the boys also learnt.

Occasionally, a search for insects can lead to something out of the ordinary and of scientific interest, such as the bat-winged fly (left). The discovery was a result of Jamie Morris's inquisitiveness, which is matched by the curiosity of young kea (below and right). Like human children, as kea chicks grow, they develop a keen sense of curiosity.

Jamie was obviously in the right place at exactly the right time; others who have gone back up to the Homer Tunnel have failed to find bat-winged flies. The other advantage of being eight (or five) is that you are lower to the ground than adults. You are closer to the rocks and shrubs where small animals conceal themselves. So there is a moral to the story: listen to kids; they may have something to say that is extremely useful. We can do no better than take a leaf out of the kea book of parenting.

Kea chicks spend over a year growing up alongside their parents. In fact, they are still in the family circle when their parents re-nest and lay eggs one year later. The chicks are boisterous and nosy, and like many a child they can drive their parents to distraction, but kea parents are long-suffering and tolerant. A key lesson for all young kea is what to eat and what to avoid. If Mum or Dad is eating it, then it must be okay. Young kea regularly take food from their parents' mouths, a habit that adults tolerate, and even encourage. Young kea learn by imitation to survive in a harsh and dangerous environment. We, too, should remember that learning by imitation helps our kids equip themselves for the future, so we should give them good taste and good example.

Over a decade before the Maud Island frog was known to science, the McMannaway children were rounding up these ancient and unique creatures for frog racing during their Sunday picnics on the island.

Both kea chicks and children seem able to make just about everything into a game. And that's how another great discovery was made by the under-12s. Way back in 1935 the Marlborough Sounds were a little-inhabited labyrinth of bays, reaches and islands. The McMannaway family farm at Deep Bay had literally been cut out of the bush. It was a tough life, and there was little flat land to play on. A special treat on Sundays was to row over to nearby Maud Island where there was a lighthouse, sheltered beaches and easy walking over open fields. Here the large family of nine children could roam free.

On one such Sunday, while fossicking around on the rocky scree in a patch of bush high up on the island, one of the McMannaways found a little frog, and before long the kids had rounded up enough of them for a frog race. Little did they know that these little creatures were real thoroughbreds of the frog world, directly descended from frogs whose lineage goes back 200 million years. Had they stayed on the island at night, they might have seen males guarding eggs in moist nests. Another remarkable feature of these frogs is that they have no tadpoles. Instead, during the weeks that the froglets take to develop, the father carries them on his back.

Many years later Bruce McMannaway recounted the frog-racing tale to Jean Bucknell, wife of the wildlife ranger on the island. The story eventually filtered out until the frog was finally 'discovered' by science in the 1950s. At first it was thought to be the same species as one called Hamilton's frog, which lived on Stephens Island in the outer Sounds. But the little racer discovered by the McMannaway kids has sprung a surprise. More recent genetic studies have revealed that it is different, and it has now been given the name *Leiopelma pakeka*, the Maud Island frog.

Opposite

Kea parents encourage and indulge their offspring. Wherever they are, from a burning rubbish tip to a colony of burrowing seabirds, chicks closely watch and learn from their foraging parents. Chicks will even steal food from their parents' beaks, an act that is accepted as a way of learning.

Punakaiki is well known for its Pancake Rocks. But in the 1940s Jack Langridge and his classmates from nearby Barrytown School drew scientist Dr Robert Falla's attention to a large burrowing seabird. They pointed out that Westland black petrels came ashore to breed in winter; their loud cries could be heard after dark when the birds returned from feeding at sea. It is remarkable that a colony of such large birds had remained unknown to science for so long.

There are few parts of New Zealand that remain truly wild, which is why visitors are drawn to the West Coast. Here they can bliss out in a primal wilderness, where forests stretch out from snowline to surf line. But very few visitors would be aware of one of the West Coast's best-kept secrets. Locals from around Barrytown knew of it, but it took a nine-year-old boy to reveal the story to the rest of the country.

Having wilderness all around them, West Coast kids are outdoor types. Back in the 1940s, however, many of them would come home to listen to the nature talks on the radio presented by Dr Robert Falla, director of Canterbury Museum. In one particular talk, Dr Falla informed his listeners that New Zealand shearwaters, or muttonbirds, nest on offshore islands in spring and raise their chicks through summer. But some of his young listeners from the West Coast thought he was wrong. On their way home from Punakaiki School in the winter, they would see large black seabirds flying in from the sea. One of the children who had been taken inland and seen their breeding colonies was Jack Langridge. The large birds were identified as petrels, first cousins of shearwaters. He had even peered into their burrows and seen their chicks — in winter.

Jack, and headmaster Mr Feasey, then wrote to Dr Falla, explaining that West Coast shearwaters were different from the others. First, they laid their eggs on the mainland, and second, they nested and raised their young in winter.

The letter aroused the avid bird scientist's interest and led him on an expedition to Barrytown, near Punakaiki, where the schoolchildren took him up to the colony to meet a bird he did not know existed. Dr Falla was surprised and impressed by the large Westland black petrel, which he named *Procellaria westlandica*. They were most certainly raising their chicks in winter and, as he discovered, very protective of their nestlings. Some came forward to attack the visitors, who retreated. It is likely that the Westland black petrels' size, and their ability to defend nests against predators like rats and stoats, are the reasons why they have survived for so long on the mainland of New Zealand.

In recognition of bringing such an important discovery to his attention, Dr Falla gave young Jack Langridge some rare eggshells and other mementos from the museum. Jack treasured his gifts, and he never forgot the day when he was able to tell a scientist something new. And though Dr Falla was one of New Zealand's pre-eminent scientists and museum directors, he always found time to answer questions and encourage young people. Just as his own interest in the natural world had been stimulated by his teacher while at primary school in Invercargill, he was more than happy to do the same for others.

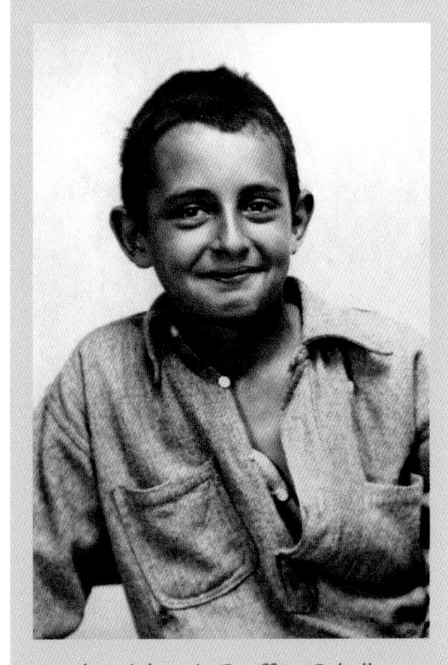

As a boy (above), Geoffrey Orbell saw an image (below right) of the supposedly extinct notornis, which we now call the takahe. It became the bird of his dreams and began a lifelong obsession, leading to his rediscovery of the bird. Fifty years later, 90-year-old Geoffrey Orbell (below) was reunited with the takahe on the shores of a lake now known as Lake Orbell (above right), which lies in Takahe Valley.

It's possible to be fascinated by something we have never seen, but would love to, and it's equally possible to be fascinated by something that is supposed not to exist, but just might do.

'Supposed' is an interesting word. It implies that the opposite might be true. 'Supposed not to exist' implies that it might. That possibility, woven in a story told by his mother, was enough for a seven-year-old boy to transform a photograph of the supposedly extinct flightless takahe into a life-long obsession that eventually led to its remarkable rediscovery.

The boy was Geoffrey Buckland Orbell from Winchester in Canterbury. He went on to study medicine at Otago University, then, after marriage and taking advanced studies in the US and UK, he established an eye, ear, nose and throat practice in Invercargill. Through all those years, however, the notornis, or takahe, had remained for Geoffrey Orbell 'the bird of my dreams'.

Takahe are large, blue flightless birds with a ruby-red beak and head plate. They are rather like oversized pukeko, to which they are related. When a takahe was caught by a dog near Te Anau in 1898, it was believed to be the last of its species. It was the same bird that young Geoffrey's mother photographed on display at the Otago Museum.

It's a wonder Dr Orbell had any time to dream of birds. Not only was he a busy medical man, he was also on several school boards, supported American Field Scholars, served as deputy mayor and was an amateur boat builder. He was involved in many Southland clubs, including sports cars, swimming, sea scouts and game fishing. He was also president of the New Zealand Deerstalkers' Association, and it was this interest that took him to Fiordland on an expedition to find a lost bird.

In April 1948, while deerstalking in the hills west of Lake Te Anau, Orbell saw a large, splayed bird footprint. He had no camera, so measured its length by making a nick on his pipe. Afterwards he consulted several experts. Some thought it was the print of a heron; others disagreed, saying it was the print of no known bird. This was the response Orbell needed, and he mounted another search.

In November, he and three companions took a boat from Te Anau back to the same valley where he had found the print. They had only that one day to explore the area and maybe shoot a deer along the way. Several hours went by, and Orbell had already begun the long walk back to the boat when he saw a familiar bush. He was on the very spot where he had earlier found the footprint. As he paused to take stock of the situation, a takahe nonchalantly walked out from behind a snow tussock bush. It was as simple as that. The bird of his dreams was real.

He cooeed his companions, and the bird remained close by. They then quietly encircled it with a fishing net that Orbell had in his pack. He'd also thought to bring a movie camera, and once the bird was in a suitable position he shot off a whole roll on this ghost from the past. The bird was unafraid and remained close to Orbell's party. As they sat and watched, another takahe stepped out from behind the snow tussock! It was obviously a pair, with the likelihood of eggs or chicks somewhere nearby. The group quickly withdrew from the scene of their remarkable encounter.

While we celebrate the fact that a bird as large and as brightly coloured as takahe was rediscovered after remaining hidden for 50 years, the story of the takahe also celebrates the imagination and curiosity of children.

Orbell's movie footage scotched any concerns that the takahe might have been extinct. The rediscovery was front-page news in New Zealand and around the world. *National Geographic* and many scientific journals wrote it up as a major discovery. The takahe was also the subject of the first ever episode of what was to become the locally produced long-running television documentary series *Wild South*.

I do not know whether curiosity kills cats. One thing I do know is that curiosity is a gift for life and it is the secret to the success of the under-12s in finding species! To be curious requires no toys — just senses; no experience — just fascination; no rules — just play; no deadlines — just time and space. There is no shortage of species waiting to be found. It is estimated that we have discovered and named only around one-tenth of all the millions of species that exist on our planet. If that's the case, we need to get busy. Perhaps it's time we unleashed the children.

Chapter 8

Giants

Fee-fi-fo-fum,
I smell the blood of an Englishman.
Be he alive or be he dead,
I'll grind his bones to make my bread.

Those four lines terrified me as a child. They came from the lips of a child-eating giant, hell-bent on hunting down young Jack, who had entered his realm by way of a magic beanstalk. My children were more fortunate: their bedtime story-book giant was far more benign, hence fewer nightmares. He was the BFG (Big Friendly Giant), from the book written by that master weaver of tales, Roald Dahl. The BFG didn't eat children — he ate snozzcumbers and drank frobscottle, causing him bouts of flatulence ('whizzpoppers'). He travelled the world at night, his huge ears detecting dreams, which he puffed into the bedrooms of sleeping children with a trumpet-like blower. Thanks to the BFG, I made peace with giants, and was later to discover that we have a surprising number of them in New Zealand.

Our BFG would have to be the kakapo. I have cradled one in my arms. It was maybe 3 kilograms, but I'm told kakapo can get up to 4 kilograms and measure over half a metre from beak to tail. That's enormous for a parrot! They became giants through a bit of evolutionary island magic. On continents, animals evolve to run or fly away from predatory mammals. On islands, if there are no mammal predators, the rules that cover running or flying from danger are considerably relaxed. In fact, islands are evolutionary laboratories with no teacher in charge. The result is some bizarre and eccentric evolutionary experiments. New Zealand was one such island group; for the longest time there was nothing to fly away from, so kakapo became supersized, cuddly, moss-green bundles of cuteness that spent most of their time on the ground. Unsurprisingly, the arrival of predatory mammals very nearly drove kakapo to extinction.

It's the human condition to think that what is happening now, what we see around us now, is normal — but it isn't, in so many ways. For example, kakapo are still hovering close to extinction, and that's far from normal. What would be normal is to have these giant cuddly parrots in our backyards, from one end of

It is thanks to dedicated people like Don Merton (below) and Gideon Climo (right) that we still have our BFG. And it is thanks to kakapo recovery teams that numbers are slowly increasing. It is now possible to meet a kakapo ambassador like Sirocco (below right) face to face. Perhaps one day, these oversized parrots might even be familiar backyard visitors.

New Zealand to the other, from coastline to snowline. Imagine kakapo as pets! That probably won't happen, but, thanks to enormous effort by DOC, they now have a good chance of recovery.

Central to that effort has been fathoming the mysteries of kakapo breeding, which doesn't happen every year. On Whenua Hou, Codfish Island, where most kakapo survivors live, DOC discovered that the key to kakapo breeding was the fruiting of rimu trees. If the season is cool and fruit does not ripen, kakapo don't breed. That may happen for three summers in a row. But, when a warm summer finally arrives, it's all on. The fruit turns from small and green to swollen and red and luscious, and kakapo clamber up into the branches of rimu trees to gorge themselves. Rimu fruit puts lead in the kakapo's pencil, and chicks in the nest.

Rimu fruit may improve a kakapo's sex drive, but it is not fruit as we know it. Apples, plums and apricots all come from flowering plants, which hide their seeds deep within the sweet flesh. But rimu, like all other conifers, has naked seeds. They are carried on the outside of the fruit, sticking out from it like a foot — hence the scientific term podocarp, Greek for 'foot fruit', which is applied to rimu and its relations.

The podocarps belong to a very ancient southern hemisphere family, and in New Zealand they attain their biggest size. Our giant podocarps — rimu, totara, miro, matai and kahikatea — are, in fact, descendants of a family of 'wooden dinosaurs'. Of these mighty trees, kahikatea is the biggest and perhaps the only one that might have existed in close to current form when real dinosaurs ruled the world.

It's hard to find the biggest of the giant wooden dinosaurs these days, as most of them have been lost to fire or felled for timber. But

It has taken some time to solve the mysteries of kakapo breeding, which doesn't happen every year. We now know that these giant parrots will raise young following a warm summer. The key to their success includes heavy fruiting of giants of another kind, podocarps such as rimu (above) and kahikatea (right).

there are some, and the biggest podocarp in New Zealand is reputed to be a kahikatea found southwest of Hamilton, hidden somewhere beneath the dark and brooding clouds that swirl around Mt Pirongia.

Not far down the track beside the Kaniwhaniwha ('dancing') Stream, I am soon among kahikatea. The giant, reputedly as high as a 14-storey building, isn't around here. These are all very young, and sexually active! Some are 'boy' trees, others are 'girls'. The boys have masses of tiny pollen-producing cones; the girls are smothered in unripe blue-green fruit that gives the whole tree a soft pastel-blue colour. It's February, and although the fruit buds were pollinated in November, they won't ripen into 'foot fruit' for another month. A kereru, or native pigeon, flies past the blue 'girls' to find something to eat that's ripe and ready.

Lit up in a patch of sunlight, a kereru's purple and green feathers and red legs glow neon. This Day-Glo bird is also a giant, a jumbo version of fruit doves found on Pacific islands. Though they consume a wide variety of fruits, kereru are the only ones that can open their bill wide enough to swallow the largest fruits of the forest. Around here, these include tawa, nikau, hinau and miro, which all seem to be ripening up.

I watch as another kereru scarfs down supplejack berries. As I move to sample a red berry on the ground, I go sprawling, thanks to supplejack, the best foot tripper in the forest. The big-chested kereru cocks its head at my fall. These giants do a great job of spreading tree seeds, but their own species isn't doing at all well. It's the same old sad story: stoats, possums and rats eat their eggs and chicks, which means declining numbers, with some local populations predicted to go extinct in the next 40 years.

Kahikatea are descendants of a family of ancient wooden dinosaurs. Forests of these mighty trees still remain in only a few remote areas in New Zealand. In such primeval forests one can imagine that dinosaurs might reappear at any moment. Other giants breed in these forests. They include kereru, or native pigeon, the only bird able to swallow and disperse the large seeds of giant fruits like karaka (right), tawa (far right), taraire and puriri, found in northern lowland forests.

We tend to overlook many of the other forest giants towering overhead, but the mamaku, or black tree fern, is one of the biggest ferns in the world. It grows to 20 metres tall with fronds that are 5 metres long. Underfoot there are giant mosses and a giant liverwort (above), which are found beside forested streams and may each cover more than a square metre of riverbank.

I realise as I lie there chomping fallen berries that the patch of light that warms me and the kereru is being used by yet another giant. Around me is a grove of mamaku, or black tree ferns, the tallest tree ferns in the world. These ferns must be 20 metres tall with 5-metre fronds, but they grow thin and gangly because they are in a race for the light with tawa and rewarewa trees. This race would have begun right after floodwaters ripped through here, taking away much of the old flood-plain forest.

My podocarp pilgrimage hits a mud wallow; but that's okay, because in the damp overhangs there are more giants that outrank the foot fruit dinosaurs by many millions of years. I bend to measure my hand against the height of the largest moss in the world, *Dawsonia superba*. With my fingers on the ground, the spore capsules reach my elbow. That's a big moss, and the secret of its size is water. As long as cells within their stems have enough water, they retain their turgor and their world-record height, but during a drought they bow down like beaten athletes.

Along the bank of Blue Bull Stream I spy another botanical Olympian. People don't tend to get dewy-eyed on first meeting the biggest liverwort in the world. It's just a fleshy green blob. But liverworts are worth another glance. The name helps tell their story. The 'liver' part is from its large, flabby lobes, rather like sheep livers in butcher shops. As for 'wort' (which simply means 'plant'), this refers to a belief, going back to the ancient Greeks, that if a plant had a shape similar to a human organ or parasite, then it was thought capable of curing diseases of that organ or that parasite, hence toothwort, lungwort and, of course, liverwort. The theory was called the Doctrine of Signatures.

On up the valley of Blue Bull Stream, the track allows glimpses of forest on the other side. The trees are mainly mid-height broadleafs such as tawa, rewarewa and mangeao. This valley was well logged over; it will be many years before podocarps overtake them. It seems unlikely that a giant could possibly survive up here. But signage claims its existence, so I continue.

As instructed, I do not cross Blue Bull Stream at the swing bridge, but keep on the same side. A 10-minute scramble later and suddenly, there it is, not right in front of me, but straight across the steeply gorged creek. It takes my breath away. It's so much bigger and broader than any other tree around it. It's not barrel-chested broad in the trunk like the giant kauri Tane Mahuta. This kahikatea is more slender, and has twin trunks, which separate a third of the way up. It's hard to tell which one attains the actual maximum height, which is a neck-cricking 66.5 metres.

It seems impossible that it has survived for so long. I don't mean it has survived when so many other podocarps have fallen to the logger's axe. I guess it was too far up in the head of the valley to be worth the logger's while. No, its survival seems impossible because it grows on the side of a steep gorge beside the river. I think it has managed to stand tall for several centuries because of the enormous buttress roots, which must radiate up and out in a massive root plate that both 'bolts' it securely to the base rock and reaches up into the leaf litter to find its nutrients.

Sitting beside this wooden dinosaur, I am happy — and sad. Happy to be here, and happy that it's still here, but sad that it is

The kauri stakes a claim to being a giant, but the tallest giant of the forest is kahikatea. Buttress roots anchor slender trees that grow up to a neck-cricking 66.5 metres. In the past, however, even mightier giants would have fallen to the logger's axe. Of all the podocarps, kahikatea suffered by growing on alluvial plains, which were cleared to become our most productive market gardens, dairy farms and racehorse studs.

such a loner. We have lost so many of our kahikatea forests.

One of the few places to find pure stands of towering kahikatea, growing as they would have when itchy dinosaurs rubbed against them, is beside rivers like the Ohinetamatea in South Westland. Many years ago, I rowed up this river with the late Geoff Park, a man who was kahikatea's greatest advocate. As Geoff said, 'Westland retains what New Zealand has lost ... the oldest and rarest of our forest trees ... we are seeing what Cook and Banks saw in 1769.'

Indeed, in 1769, James Cook and Joseph Banks marvelled at some fine kahikatea while rowing up the Waihou River near Thames. Cook commented that they were '90 feet to the first branch ... sticks of first rate quality'. He was assessing the trees as timber for spars or masts for the ships that would keep Britannia ruling the waves.

Of all the five podocarps, kahikatea has paid the ultimate price for living in the wrong place. Kahikatea are the podocarps of alluvial flat land and fertile plains. They grew where cities like Hamilton and Lower Hutt sprung up, their tall crowns dominating land that would soon become fields of highly productive market gardens, dairy farms and racehorse studs.

So, it's a big thank you to South Westland, Mt Pirongia and the pockets of kahikatea forest around New Zealand. Other members of the family of giant wooden dinosaurs — totara, rimu, miro and matai — might be doing better, but reflect on this: it's very lucky that we have any dinosaur forests at all in New Zealand.

Had a truly dangerous and appalling giant stomped on New Zealand 65 million years ago, we could have lost the lot. Instead, it stomped on the northern hemisphere. In his book *The Eternal Frontier*, Tim Flannery describes how the giant meteor that slammed into Earth, contributing to the extinction of the dinosaurs, had an equally devastating effect on dinosaur forests above the equator.

But scientists have worked out that this meteor of malevolence, over a kilometre in diameter and probably shaken loose from the asteroid belt, must have approached Earth at a low trajectory from the south; most likely it passed just over the crowns of our vast podocarp forests. Our giant weta and snails and tuatara would have seen its fiery track close overhead.

Instead of hitting us, it punched a 5-kilometre hole in the planet's crust, close to Mexico's Yucatán Peninsula. The impact blasted many thousands of times its mass of material into the atmosphere and back into space. The fallout of erupted material blackened skies for months, depriving the northern hemisphere of the sun's life-giving energy. Temperatures dropped to near freezing, dinosaurs and other animals suffered catastrophic losses. Northern hemisphere evergreen dinosaur forests, similar to our podocarp forests, fared worse than deciduous trees. Unlike the deciduous trees, which lose leaves in winter, evergreens had no way of shutting down during the months of cold and dark. As a consequence, there was a mass die-off of these evergreen forests, which became extinct across most of the northern hemisphere.

But what else was lost when this extra-terrestrial golf ball landed on the planet? Maybe the giant meteor was a killer of giants. Maybe those lost forests were home to even bigger ferns and trees and liverworts and other botanical wonders. Maybe the giants that we see up any track in any New Zealand forest had even bigger northern hemisphere counterparts. Who knows?

Sixty-five million years ago, Earth was hit by a giant meteorite, which punched a hole in the planet's crust 5 kilometres across. The fallout from this impact blackened skies and masked the sun's life-giving rays, cooling the climate and causing a mass die-off of northern evergreen forests. Was it also a killer of northern giants? While tuatara and weta may have seen the comet streak overhead, they and ancient New Zealand remained largely unaffected by its impact and aftermath.

Chapter 9

Waddling Mouse or Boxing Bat?

We live in an age of instant information that has shrunk the world down inside our mobile phones. Having all this information at our fingertips is brilliant, wonderful and empowering. Of course, a mass of information has a downside: confusion, distraction, trivialisation and misinformation. Messages and media swirl around, coming from all sides, blurring perspective and bludgeoning reason, leaving us dazed and confused. But there is an antidote to information overload: detail. Some say 'the devil is in the detail'; not so. In detail we find the gods of truth, knowledge and understanding.

Examining anything in detail requires time, sometimes lots of it. But the rewards can be immense. Incredible discoveries are possible; mysterious worlds can be revealed. That's exactly what happened in Central Otago. From a hole in the ground, on a riverbank near St Bathans, scientists have discovered a portal into New Zealand's past. In the course of 10 years and countless hours of painstaking work, fossil bones of over 40 animals have been discovered. From fossils, scientists have built up a picture of what southern New Zealand looked like 20 million years ago. It is a New Zealand that you would not recognise.

In the early 1990s, from a hole in the ground near St Bathans in Central Otago, scientists began uncovering a world that existed here 20 million years ago. It is a world you wouldn't recognise. The climate was tropical, with palm trees; unrecognisable yet warmth-loving reptiles, birds and fish lived in and around a vast lake 10 times the size of Lake Taupo.

Picture this. You are in Central Otago, but it's totally unfamiliar. There are no recognisable landmarks. The weather is hot and very humid. You are in a forest of tropical trees and shrubs, and you are following a loud noise made by many birds. The sound takes you out into an area of reeds and rushes. Now you can see the birds in the distance; great flocks of waterfowl and waders twist and spiral in clear morning light. The noise is tremendous; there must be thousands of them. You move closer. As you move through the sedges, you narrowly avoid standing on eggs in a nest concealed among the reeds. Suddenly, a large bird with a massive, clacking bill confronts you. It is like a weka, but much bigger.

Putting distance between you and the strange bird brings you to the shore of a lake. It's obvious now what has disturbed the flocks of waterfowl — several large crocodiles are basking on their roosting beach. Beyond the shore, flocks of flamingo-like birds step through the shallows. What is this mysterious place?

If you could rise above the lake, you would be astonished. From a 'Google Maps' perspective you would see that the lake is more like an inland sea that stretches from South Canterbury all the way down to Southland; from coastal Otago in the east over to where the Southern Alps should be in the west — it must be 10 times the size of Lake Taupo. You are looking at southern New Zealand as it existed 20 million years ago.

Evidence of this extraordinary lake and the strange and unfamiliar animals that lived around its edge has all come from the hole in the ground mentioned earlier. The hole is the size of a small suburban section, yet from it a treasure chest of fossil evidence has been unearthed that has enabled scientists to build up a detailed and precise picture of this unexpected and unknown lost world.

The fossils all come from a gently sloping layer of soft rock 2 metres below the surrounding paddock. This layer was once the bed of the vast ancient lake. Today, down in the hole, researchers have just finished the grunt-work of removing clay overburden to expose another section of the dark layer of ancient lake bed. 'Fish,' says leader Trevor Worthy, as he nudges yet another fossil bone fragment from the soft mudstone 'fudge'. Fishbones represent 99 per cent of the bones discovered, but he and the group work energetically, hoping that the hundredth bone will be the jackpot. They have struck jackpots before.

It was the chance discovery of a crocodile bone that started a wider search for fossils. Since they began exploring the site in the 1990s, scientists have been able to describe in detail what lived here and what kind of climate could support crocodiles, turtles, flamingo-like birds and other unexpected creatures.

The wealth of fossils at the St Bathans site has drawn scientists from throughout Australasia. Dr Steve Salisbury (above), a crocodile scientist from Queensland, has discovered that our crocodile was related to island-dwelling crocodiles once widespread throughout the Pacific. Other scientists (above right) are looking for the bones of birds, reptiles and mammals; but 99 per cent of their finds are fish bones. These reveal that there was a greater variety of galaxiids than exist now, including three species larger than our giant kokopu (right) along with other fish groups that are now extinct, such as carp and grayling.

Among all the fish bones, there is the occasional special prize. Not only did the St Bathans fauna include crocodiles, but giant horned turtles, genus *Meiolania* (above; skeleton), once lived here too. Like the crocodiles, these turtles were once widespread on islands in the Pacific. Today, they are all extinct, and the closest we can get to imagining these spectacular reptiles would be to visit the giant tortoises of the Seychelles or the Galápagos Islands (left).

These days the river flats are covered in frost or snow for most of winter, but 20 million years ago, when these ancient animals lived, average temperatures were 7 degrees warmer — similar to Northland or South Queensland, making this lake a subtropical wonderland. How come? It can partly be explained by the movement of continents. Around this time, warm subtropical waters expanded southwards and swirled around ancient New Zealand. This was possibly due to reduced ice volume in East Antarctica, and also because the rising of the Indonesian archipelago redirected warm tropical ocean currents to flow southwards into the Tasman.

Other possible climate-changing events around this time were the opening up of the Drake Passage between South America and the Antarctic Peninsula, and the recent separation of Australia from Antarctica. The circumpolar current had just begun flowing and Antarctica was just beginning to cool into an ice continent. Back then it still had forests, and southern New Zealand was still nice and warm.

The hole in the ground near St Bathans has become one of New Zealand's most important portals into its warm past. And as scientists discovered a few years ago, some very strange

inhabitants lived way back then. One of the strangest was a land mammal that caused a lot of interest — not merely because it was a land mammal, but also because of its gait, which earned it the name of 'waddling mouse'. Most four-legged animals have their legs attached underneath the body. The legs of this weird mammal seemed to be attached out to the side, like a lizard or a crocodile, which would have meant that instead of walking, it would have waddled.

The discovery of the waddling mouse caused huge interest for another reason. It was the first real evidence that New Zealand once had its own bona fide land mammal. Up until then, it was thought that the first land mammals to inhabit New Zealand were rodents called kiore that came with Polynesian settlers less than a thousand years ago. So, if the fossilised bones did belong to a 20-million-year-old land mammal, it changes our whole nature narrative. Up until now, we have been convinced that New Zealand had no land mammals before humans came, and that birds dominated every ecosystem and niche. A lack of mammalian competition meant that birds could evolve into all manner of shapes, sizes and 'occupations'. So what part did the waddling mouse play in the story and what happened to it?

Kiore, the Polynesian rat (below left), was thought to be the first land mammal to inhabit New Zealand, arriving less than a thousand years ago. The discovery of 20-million-year-old land mammal bones at St Bathans caused great excitement. But what exactly was it? From the shape of the hip bone, Trevor Worthy (below) and others thought it had a clumsy gait, and it was named the waddling mouse. Was it a small, primitive mouse-like animal, perhaps similar in appearance to the Australian antechinus (centre left)? Or does Australian bat scientist Suzanne Hand (below right) have the answer? She wonders if it might have been a bat that relied heavily on its hind legs, as New Zealand short-tailed bats (upper left) do today.

It was Trevor Worthy, an eminent vertebrate palaeontologist, and the man down the hole the day I visited, who with several colleagues first described the waddling mouse. One spin-off of the discovery was that it brought more interest, more funding and more scientists to the excavations in Central Otago. One of those scientists was Australian bat expert Suzanne Hand, who looked at the bones of the waddling mouse and thought she saw a quite different creature.

Animals exist in three dimensions, and sometimes examining fossil bones by looking at them in another framework allows researchers to build rather different pictures. Remember the early nineteenth-century illustrations of early moa, kiwi and penguins? The mainly English artists had only flattened bird skins and piles of bones as guides to work from when sketching and painting New Zealand's strange and exotic bird species. Some results were hilarious, others cannily accurate. You do your best with what you've got. That's how Trevor and his co-authors came to see what became known as the waddling mouse. But turn the pelvis 90 degrees, as Sue Hand did, and you have a mammal that stood on hind legs.

Sue has spent several summers with the team at St Bathans. She, like everyone else, shared the tremendous excitement of the discovery of a 'new' mammal, but later, when more complete specimens were found, she and the team took another look at the fossil femur (upper leg bone). In her mind she rotated the pelvis into which the leg would insert, and she saw a strange but familiar bat. She remembered watching video footage of these bats and being spellbound at seeing them standing, in fact sparring with one another in exactly that same stance.

Sue had also seen South American vampire bats standing toe to toe and boxing with one another, possibly in displays of dominance. Standing could explain the fossil femur's strange angle of insertion into the pelvis. So what was this mammal? Was it a waddling mouse, as Trevor had first thought, or a boxing bat, as Sue was suggesting, or was it something else altogether? The answer is that researchers still don't know.

The truth may yet be revealed not in the bones of its pelvis and leg, but in its teeth. A number of identifying characteristics separate bat teeth from those of other mammals. If they had a tooth, particularly a molar, Trevor Worthy and Sue Hand could quickly tell whether the mystery mammal was a bat or a mouse-like creature. Thus far, researchers have found a few random bones, a single tooth (that is not from a bat) and some jaw fragments. As no molar teeth have yet been found, this animal's identity remains a mystery. But the clue to solving that mystery could already be in their hands.

Each summer, over more than 10 years, as researchers have carefully expanded the hole and painstakingly picked over the ancient lake-bed fossils, they have sorted and bagged every bone, shell fragment and piece of sediment from this special layer. After each summer's dig, these bags are sent to Adelaide, where over the following 12 months the material is analysed and species are described. Maybe the mammal's tooth is already sitting in one of those bags. And if the tooth is not there, it may be discovered next year, or the year after, or who knows when.

In this kind of research, bigger bones get instant attention. A few days earlier, members of the team pulled a fossil bird

Each summer, the scientists at St Bathans gather hundreds of bags of fossil-rich material (above), which is sent for painstaking analysis and identification of the smaller fragments. Sometimes it is difficult to know exactly what the animal looked like. Early scientific illustrators had similar problems depicting our kiwi (above left) when working only from preserved skins. Fortunately, some fragments are instantly recognisable. Moa eggshell (left) proves that moa lived here 20 million years ago. And fragments of bone reveal that some kiwi were half the size of today's little spotted kiwi (right).

leg bone out of the riverbank a little way upstream from the site. The bird experts identified it as belonging to an adult kiwi, but what was surprising was its small size. They reckoned that it would have been half the size of the smallest living kiwi, the little spotted kiwi. Imagine it — a kiwi not much bigger than a tui!

Mammal mystery aside, the discovery of the kiwi bone, tuatara bones, moa eggshell and bones of many other species that lived around that ancient lake provides compelling evidence that many modern species or their close relatives already existed 20 million years ago.

This date also seems to have cast doubt on another recent theory about our ancient past. As mentioned in chapter 6, it is now agreed that New Zealand is all that exists of a much larger subcontinent called Zealandia that lies beneath the sea. A contentious theory that has been advanced recently states that the entire Zealandia land mass sank beneath the waves for several million years, and that later on the bit we now call New Zealand was thrust up by immense tectonic pressure. However, the dating of the thousands of bones from St Bathans shows

Summer temperatures in Central Otago today reach into the 30s, driving many of us into the Manuherikia River (above), close to the fossil site. Here the riverbanks are full of fossil deposits of fabulous creatures from that far-off tropical time. The adzebill, genus *Aptornis* (above right), was like a giant weka on steroids. Did ancient, primitive frogs brood their young on their backs as male Archey's frogs (below) do today?

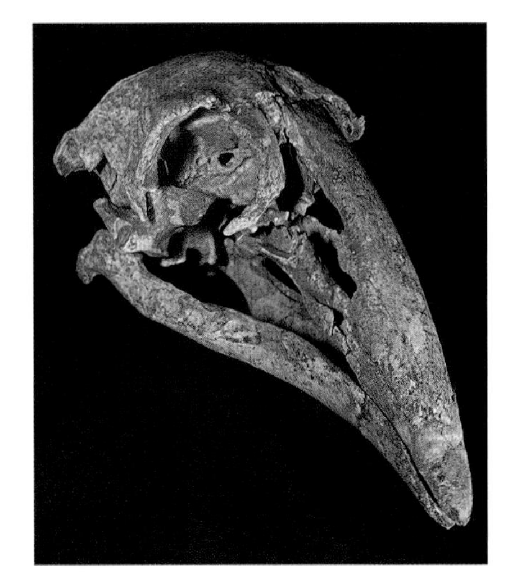

New Zealand had a rich and diverse fauna only about three million years after the point when, according to the theory, the land mass should have been deep under water.

Many of these animal species, such as leiopelmatid frogs, tuatara and several New Zealand genera of birds, are endemic (known nowhere else). It is most unlikely that, in the short space of time between the 'drowning' and the known date of when these fossil animals lived, they could all have dispersed, colonised this land mass and evolved into such special New Zealand species, so different from those elsewhere in the world.

Moreover, from the St Bathans dig we know that all the really special animals (kiwi, tuatara, moa, and so on) were already present 20 million years ago. Many of these had already undergone many millions of years of evolution in Zealandia and some ancestors undoubtedly were aboard when the sub-continent separated from Australia between 80 and 60 million years ago. The Central Otago portal into the past has helped sink the theory that all of New Zealand was once under water.

For the scientists, tantalising hopes abound. Will they find the serrated-toothed bill of our extinct merganser (below), or the fragile wing bones of the flightless Lyall's wren (below right)?

The age of the fossils at St Bathans points to land having been here for a long time before, during and after the supposed total sinking. What's more, geologists can tell that the land the animals occupied around the lake covered a large area. From the spread of ancient river gravels in the lake, they estimate the largest rivers draining into the lake were almost a kilometre wide. Such large rivers could only have existed if there were significant hills, even mountains, nearby to supply all the gravels transported by the lake's tributaries.

Scientists estimate that ancient Lake Manuherikia, named after the river that runs close to the fossil site, covered 5600 square kilometres; that's 10 times the size of Lake Taupo, or as big as all the existing lakes in New Zealand combined. How do we know this? Ironically, it was the search for coal in Central Otago that first drew attention to the size of the ancient lake. Back in the 1970s, palaeogeographers discovered evidence of lignite coal in several locations in Otago and Southland. The thing about lignite is that it is soft, low-grade coal. It is formed not from fossil forests, like hard coal, but from reeds and other vegetation that once grew in swampy deltas, usually where rivers entered ancient lakes. The extent of the lignite deposits pointed to there being extensive reed beds, indicating that huge rivers once drained into one vast lake.

The hole that Trevor and his team are exploring was from the shallows of this mega-lake. These shallows would have been the kind of place where crocodiles rested and birds roosted. Gulls, herons, 'flamingos', hawks and eagles would have regurgitated and defecated bone, shell and other material. But most of the bones are typically smaller than 20 millimetres, and most likely were washed and sorted by currents that swirled close to the shore, and rolled about in the shallows. Of course, some bones were of animals that lived and died in nearby forests, and many of their bones would have been washed into the lake during floods.

Vast braided rivers, similar to today's Upper Waimakariri (preceding pages), flowed into the ancient mega-lake at St Bathans. Lignite deposits in Otago and Southland indicate that expansive reed beds (left) would have spread around the edge of this lake. In these warm shallows, crocodiles and vast flocks of waterbirds would have thrived. It is so different to the dusty, gravelly riverbed where the scientists work today (above). But had this site not been discovered, we could never have peeked through this keyhole into the ancient world that once existed here.

Visiting the team at St Bathans was an extraordinary experience, but what amazed me was that the small area of lake bed exposed in this one small hole has allowed researchers to peek through a keyhole into another world. And this narrow view is almost unique: there are very few other sites anywhere in New Zealand that can further expand the picture of life in ancient times. It is fair to say that without the St Bathans site, and more than a decade of painstaking and detailed work by an A team of scientists, we would know practically nothing of a very different New Zealand that existed around 20 million years ago.

Chapter 10

Playing God

I have played God many times — we all have. Rimu forest once flourished where my lettuces grow, but loggers, then farmers, then several owners of my property and I have transformed it into lawn and vegetable garden, fringed with rhododendrons, gums, bamboo and a giant copper beech. There are native plants, some from nursery seedlings and others by kind courtesy of bird droppings. In my world, and in yours, I suspect, we are increasingly taking over nature's work. On a global scale, we up the ante by moving mountains, modifying genes, changing the climate, bringing water to the desert, transforming vast areas of forest into pasture and saving and destroying other species. We are staking a claim to be in charge.

In New Zealand, we began pushing nature aside as soon as the first hillside was set ablaze, sending fleeing moa and other birds towards waiting hunters. At the height of our burning frenzy in the late nineteenth century, we were clearing forest as fast as Amazon rainforests are being razed today. But we have saved forests too. The wilderness movement began with the establishment of Yellowstone National Park in the 1870s, and

As soon as we push nature aside we play God with the environment, whether it is setting fire to scrub (below) prior to grazing or importing pasture seed (and with it, weeds). The result is that some landscapes look more like a Monet painting (far right) than anything we might recognise as New Zealand. Sometimes we don't foresee complications until it's too late. White butterflies (right) probably stowed away on imported cabbages. And although house sparrows (below right) were brought here to control insect pests, they quickly became grain-stealing pests.

by 1887 we had our first New Zealand national park, Tongariro. Now we have 14. However, by protecting areas of scenic and natural beauty in national parks, we are playing God. We are dividing our country: some for nature, the rest for us.

We gave nature another kick in the pants by bringing in foreign species. The first lot was brought by Polynesian voyagers, and then European colonists brought a whole lot more. Some species, such as blackbirds and thrushes, were reminders of home. Others, like salmon, trout, deer, rabbits, moose and chamois, came from a desire to turn this land into a hunting and fishing paradise. As each new import became established, it usually harmed a local species; the worst harm was done by predators. The list of introduced species is so long, it has given New Zealand the dubious record of having more introduced species than any other place on Earth. But that is starting to change. A small group of New Zealanders is turning the tide against the most destructive of introduced species. They kill to restore life; and in 2001 they made a killing on a scale never seen before anywhere in the world.

European settlers envisaged subantarctic Campbell Island would offer ideal farmland for sheep and cattle, but the harsh climate soon defeated the farmers, who deserted the island, leaving their livestock, their house cats and their camp followers, rats. In succeeding years the cats died out and livestock were removed; only rats remained, building up in numbers as high as anywhere in the world and causing devastation among wildlife. Working out how to rid the 11,000-hectare island of rats would be a huge challenge for project manager Peter McClelland. Some thought it impossible.

A chopper rises up and over a ridge hugging the contours of the slope. It rains death on those below. This is no war movie: this is war. In 2001, cold, wet and windswept Campbell Island, partway between New Zealand and Antarctica, became the setting for the largest island rat eradication programme ever attempted in the world at that time.

Norway (or brown) rats are, next to humans, the most successful mammals on the planet. They have been on this remote island since the early nineteenth century, when they stole ashore from sealing or whaling boats. When naturalists arrived a few years later, they found only the larger seabirds; rats had wiped out every last land bird as well as the smaller seabirds.

Cows, sheep and cats were left behind from failed attempts at farming this windswept outpost. Cats died out naturally, but restoration of this nature reserve began when DOC removed sheep and then cattle. Rats remained on the island in numbers estimated as being higher than any other place in the world. Removing them would be an extremely tough challenge; many said it would be impossible.

For project manager Peter McClelland, the list of challenges was enormous. Not only was he responsible for safely transporting people, helicopters, supplies and poison to one of the most remote locations on the planet, but then he had to oversee a military-style operation of such precision that afterwards no rat could be left alive anywhere on the 11,000-hectare island. The operation took place in winter, when rats were hungriest, their numbers lowest and there was the least chance of young surviving. Also, in winter, most breeding seabirds are absent, removing them from any risk of poisoning and also removing a potential alternative food source for the rats.

Peter divided the island into four sectors, each marked into a GPS-guided course of 40-metre swathes for helicopters to follow. In this way, they peppered the entire island with poison, even landing pellets on narrow cliff ledges. The campaign was based on a single poison drop.

The challenge for Operation Campbell Island was to spread enough poison in one drop to kill every last rat (above left). That meant ensuring that every rat on the island was within a short walk of a block of poison (above). The frustration lay in having to leave the island (right) without knowing whether the campaign had been successful or not.

Among rats all over the island, the message clearly went out that there was food. It was good, tasting like the seed heads of tussock, and there was lots of it. In Norway rat hierarchy, big males are the first to have their fill. Lesser males, females and young rats hold back until the boss has feasted and then stored enough under the tussocks for later. But the boss males wouldn't get back to their stores, as the poison, lethal even from a single feed, quickly began working its deadly magic.

The poison was brodifacoum, an anticoagulant that prevents blood clotting in animals that possess haemoglobin. Affected animals die slowly but painlessly as they haemorrhage internally. The alpha rats were becoming drowsy. A few had already gone to sleep permanently, which was the signal for others to make a grab for bait on the ground and raid the storage caches. There was plenty for all.

The uncertainty for Peter and his team was knowing if a single poison drop of 6 kilograms per hectare was enough to kill every last rat, from the biggest to the smallest, the boldest to the most cautious. Their frustration lay in having to leave the island before they were certain of a 100 per cent kill. Even as they loaded supplies, equipment and personnel aboard the support ship *Jenka*, and chopper pilots were filling long-range fuel tanks for their flight back to New Zealand, they did not know if the job was finished or not.

A miracle on Campbell Island

Nineteenth-century naturalists reported that on Campbell Island all land birds had been wiped out by rats. However, they and other occasional visitors weren't to know that there were some survivors hiding away.

In 1975, a tiny population of endemic flightless ducks was discovered on a rocky islet named Dent Island. Dent Island was a safe sanctuary for the ducks, because it was 2 kilometres west of the main island and beyond the reach of swimming Norway rats. That's how a few of these plucky little ducks had survived for 150 years. Campbell Island teal are semi-nocturnal and, unlike most other ducks in the world, they live in and around seabird burrows above sheer cliffs, which denies them access to the coast. Life is tough, and it is thought that in winter their food is scarce and supplemented by insects and other invertebrates that breed in seabird guano and regurgitations.

Campbell Island teal hold a world record, unfortunately for the wrong reasons. For all their tenacity, they are listed on the IUCN Red List of Threatened Species as the rarest ducks in the world. There were only about 30 of them when discovered, but, since then, captive breeding programmes have helped boost total numbers to close on a hundred. They were also reintroduced to Campbell Island in 2004–06 and are now widespread there.

If that wasn't enough, in the late 1990s, on Jacquemart Island, a 19-hectare, tussock-covered rock lying 1 kilometre to the south of Campbell Island, a small population of endemic snipe, possibly as few as a dozen, was discovered. Since the elimination of rats, the snipe have self-introduced back onto the main Campbell Island and seem to be doing very nicely.

With the successful eradication of rats from Campbell Island, birds once again began to rule the roost. A tiny population of flightless teal (above), marooned on an offshore rock stack, was returned to the main island and these little ducks are now widespread. Many seabird colonies such as this mollymawk colony (right), with its large downy chicks on chimney-pot nests, began to flourish. What was seen for a long time as an impossibility has became a modern miracle.

New Zealand leads the way in predator eradication on islands. It began with government hunting parties eradicating pigs and goats from offshore islands, such as this group on Cuvier Island. A few years later, in 1964, two of these hunters, Don Merton (front row, second from right) and Brian Bell (front row, centre), faced a villain that was not so easily beaten: rats. When Bell and Merton went to Big South Cape Island off Stewart Island (below left), the rats were at plague level and they were forced to rescue and transfer the few surviving saddleback, snipe and wrens to nearby rat-free islands. Only the saddleback (see page 16) was saved. The snipe (above) and the bush wren (page 61, lower) became extinct, along with the greater short-tailed bat (below), which was not recognised as a separate species at that time.

The island once more became the plaything of the winds, a punchbag for winter storms. Now, too, it was a killing field for a million rats. If rats survived, it would confirm what many people believed: that this predator that we have taken to the four corners of the Earth is too powerful an adversary to be defeated. But, on the positive side, if all Campbell Island rats had been killed, it would be hugely empowering, and encourage others to attempt similar operations on an even larger scale around the world.

The Campbell Island operation had only been possible because of what others had done before. It started with the late Don Merton, or 'Mertie', New Zealand's best-known conservationist. Mertie's work in rescuing black robins from the brink of extinction became widely known around the world, as did his innovative techniques in helping turn around the fortunes of kakapo. Less well known, though, was his reputation as a killer. As a young officer with the Wildlife Service (as DOC was formerly known), this passionate bird and animal lover found himself looking at wildlife down the sights of his rifle, but it was wildlife that didn't belong. A major part of his job was shooting animals that were overrunning fragile island ecologies. Goats and rabbits, food for castaways of an earlier generation, had multiplied to become forest destroyers; abandoned cats transformed to mini tigers with maxi appetites for birds. But the villain that couldn't be beaten was the rat.

Mertie had watched helplessly as rats tore through the birds of Big South Cape Island off Stewart Island. Back then 'field' people like Mertie and herpetologist Tony Whitaker were reporting that rats were the major problem for endangered species on offshore islands. But some scientists refused to accept their evidence, citing the fact that rats could live alongside tuatara on offshore islands. Tony took issue, pointing out that on these islands the tuatara were old, some over 100 years, and that there were no young ones being recruited into the population. Rats were taking all eggs and killing all young. These populations were doomed. Mertie, Tony and all who had seen with their own eyes what was happening on islands knew that rats were major villains.

Attitudes were to change, and one key event occurred when Mertie was sent to Maria Island, a small dot off Waiheke Island. He was responding to a report that the island's white-faced storm petrel colonies were being decimated by rats. With the help of a local schoolteacher, they trapped the culprits, black or ship rats. Then they returned during the next couple of petrel breeding seasons and ringed the colony with poison, then departed, hoping for the best.

A decade later, scientists surveying Maria and adjacent islands made the astonishing discovery that they were all rat free. It seemed that Mertie and his helpers, with a bit of poison and a lot of hope, had eradicated the rats. They had achieved what many thought impossible. The news that rats could be eradicated from islands inspired others to believe that killing rats and restoring the battered ecology of much larger islands was not only possible, it had to be done.

The next rat-busting campaign was so tough it was called the Battle for Breaksea. Rugged Breaksea Island, at the entrance of Breaksea Sound in Fiordland, had been home to Norway rats for 150 years. Many native species of bird, reptile and insect had disappeared. During the winter of 1988, a team arrived determined to end the rat menace. Led by Bruce Thomas and Rowley Taylor, they cut a network of tracks across the steep, forested 170-hectare island and laid out over 700 bait stations with a new-generation anticoagulant poison. For three weeks they monitored and replenished eaten bait. Again, however, the team had to leave not knowing if they had succeeded.

After a few nervous years of regular checks for any sign of rats, it was finally announced that the Battle for Breaksea had been won. Since then, many species have returned to the island. Some birds have flown across from nearby rat-free islands; other birds, lizards and insects that were once native have been brought back to help the rebirth of Breaksea. It is now a showcase conservation island and has given impetus to the rescue of other islands suffering from death by rat.

Over the next 15 years, 40 islands were cleared of rats, including Cuvier and Tiritiri Matangi islands in the Hauraki Gulf, and Mana and Kapiti near Wellington. But what of Campbell Island? It is five times larger than Kapiti, site of the previous

Though attitudes and techniques have changed, the results are always the same. When islands are cleared of rats, the indigenous fauna comes back. Giant flightless weevils like the olearia weevil (above) and tangles of Fiordland skinks (below) reappeared on Breaksea Island in a surprisingly short time. The white-faced storm petrel (right) was one of the first birds ever to benefit from rat eradication when poison was laid on Maria Island in the Hauraki Gulf in the late 1950s.

largest successful island campaign. Could we add Campbell Island to the list of successes?

The problem in monitoring Campbell Island for rats is that few people go there. The year after the poisoning operation, the island was checked for rats — gnaw sticks soaked in vegetable oil were left out. This was repeated in 2003. Finally, in 2006, Campbell Island was officially declared rat free.

We should dance a jig and raise a glass to all of New Zealand's rat-free islands, because what happened next is pure magic. Over the years since the removal of predators, these broken, beaten island ecosystems have been resurrected. They show us something we cannot see anywhere else on the planet. On these islands, we can wander through a wonderful world where birds literally rule the roost. They offer snapshots of a New Zealand that existed before we blundered onto the scene. And now, rat-free offshore islands are among the few places where these unique old New Zealanders can be truly safe, secure and left to get on with life.

Although they have given us the immense satisfaction of having rebuilt lost habitats, rat eradication operations have their critics. The main issue is whether animals suffer. They do; there's no way round it. A bullet and a trap can be painful, and poison can sometimes kill species for which it was not intended. This is tragic and project leaders try their best to minimise accidental death. But the end is to make the habitat fit for the species that evolved there. They have nowhere else to go. We were the ones who, whether by ignorance or accident, threw these island worlds into turmoil. It's now our job to fix it.

We have been playing God on this planet for quite some time, often not in a good way, but the island rat busters from New Zealand have shown us that with the right tools, planning and belief, we can fix our blunders. Following the success of Campbell Island and other recent large-scale rat eradications, there are some who are dreaming much bigger dreams. They dream of a predator-free New Zealand. What's more, they believe it can be done.

Wherever rats have got ashore on oceanic islands, they cause havoc among indigenous species. On Campbell Island, rats exterminated the local parakeet, and forced many others like snipe, ducks and pipits to take refuge on offshore rock stacks. But the success of eradicating rats from such a huge and rugged island as this makes us realise that there is often a solution to our ecological blunders.

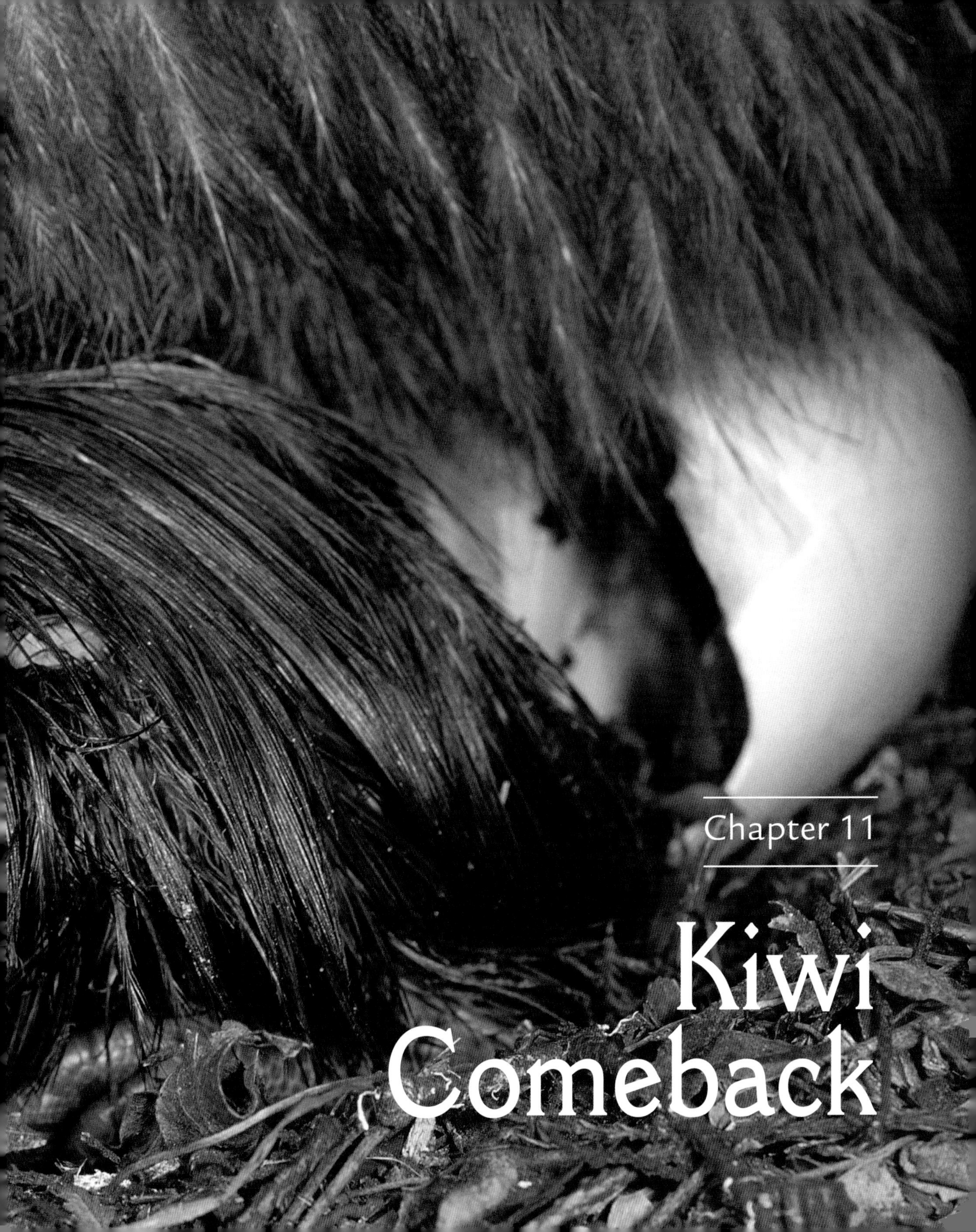

Chapter 11

Kiwi Comeback

'I am a Kiwi; but what am I going to call myself if the kiwi are all gone?'

— Philip, Whangarei Heads

After enormous effort, the chick manages to peck and push its way out of the oversized egg. Father is in attendance, as he has been for 70 days since the egg was laid. During all that time, Mother has been getting a well-deserved rest after carrying, then laying, the enormous egg, which, when inside her, took up 20 per cent of her body, pushing her internal organs out of the way.

Apart from Dad, who now closely inspects this small, gooey being, there are others who are intensely interested in the newborn kiwi chick. Somewhere in this valley near Whangarei Heads, a killer is on the prowl: a stoat.

From the moment it hatches, the chick is able to walk, albeit rather wobbly-legged at first. For the first couple of days, the chick stays in the nest, close to Dad. It feeds off the last of the egg yolk in its stomach. By day 3, hunger sends it outside, on its own, to fossick for insects among the leaves. Later, back home, it sleeps. This is the pattern for the next few days. The nest is being watched and soon the watcher makes a move.

It's the right time to do it, because a stoat is also likely to be monitoring the chick's progress. Several days earlier, kiwi handler Todd Hamilton noted a change in the pattern of movement shown by the data stream from the kiwi dad's movement-sensitive transmitter. Previously, while incubating the egg, the kiwi dad's movements were as regular as clockwork. Each evening, he would leave the nest to feed for a few hours, then return. Then one day, data showed he had stopped moving, which meant the egg was hatching.

A kiwi chick spends 70 days in an enormous egg. Such a long incubation means that from the moment it hatches, it is able to walk, and within a few days it is foraging for itself, fossicking for insects among the leaves.

The stoat — a killing machine

Whatever well-intentioned soul brought stoats to New Zealand did this country's wild creatures a monumental disservice. The damage done by stoats (and to a lesser degree by ferrets and weasels) is nothing short of catastrophic. Their efficiency as predators is due to robot-like behaviour and physiology. Here are some of the statistics of this amazingly adapted killing machine.

Physical attributes

A stoat is very thin, and although its pipe-cleaner shape is ideal for going down holes, it makes the animal vulnerable to cold. To survive, the stoat must eat up to one-third of its body weight in food per day.

Average body length 260–280 mm.

Average tail length 90–100 mm.

Heart rate Up to 500 beats per minute.

Speed Very fast; a stoat can cover ground almost as quickly as a bird can fly over it. It can leap up to 50 cm into the air, climb trees, and swim up to 2 km.

Senses

Hearing Good.

Sight Very good.

Smell Extremely good — 'odorous' kiwi are prime targets for stoats.

Predatory behaviour

Ferocity Extraordinary; a stoat will take on prey much larger than itself.

Appetite Eats six or more meals a day, but kills regardless of appetite. In cold countries, stoats store food, but in New Zealand they kill and move on.

Defences

Wariness Trappers have only one chance to catch a stoat; if they fail, it becomes extremely wary and impossible to trap. New technology to control stoats and manage kiwi is in development, and it is hoped that the job of ridding large areas of New Zealand of stoats will soon get easier and less costly.

Breeding

A female stoat has the ability to carry fertilised eggs inside her body from mating in summer until the following spring. Female kittens can be mated while still in the nest.

Todd reaches into the nest and gently scoops up the warm, drowsy chick. He has beaten the predator to this prize. He records its weight and bill length — females have a longer bill. He inserts an identity chip under the skin and records the date, 22 February. This day is also the anniversary of the terrible earthquake down south. So he gives this kiwi chick a very special name: Christchurch.

A few years ago there were hardly any kiwi in the Whangarei Heads area (below left); now there are over 300 and the population is mushrooming. This is due to the efforts of the community-led organisation Backyard Kiwi. Once every six months, project manager and kiwi handler Todd Hamilton replaces transmitters on the legs of male kiwi (above far left). This one will soon be sitting on a nest, and information from the tracking device enables Todd to monitor the bird's breeding progress from a distance and without disturbance. Distinctive local road signs (below) warn motorists that kiwi are now far more likely to be seen on these rural roads than possums!

It's February; Northland is as green as can be. Pohutukawa are past blooming but have oodles of leaf growth, and grass (and weeds) are rampant. It's such a contrast to the two previous years of severe drought. Back then, the ground was so hard, kiwi couldn't push their long bills into it to get food. They went hungry, lost a lot of weight and failed to breed. But this year, kiwi are popping out eggs all over the place. Christchurch was the eighteenth chick that Todd had scooped since September.

He traps predators and manages kiwi affairs for several land care groups that have demonstrated their desire to help the birds by forming an organisation called Backyard Kiwi. 'Backyard', because the groups were not just farmers but a diverse mix of bach owners, lifestylers and retirees living in the villages and settlements that nestle in the bays out towards Bream Head, on the north side of Whangarei Heads. As Todd tells it, 'a few years ago, there were hardly any kiwi here, but now numbers are mushrooming'. And many of them live literally in people's backyards.

Adult kiwi can defend themselves against stoats, but they are helpless against an all too common enemy in Northland: dogs. And why are dogs so harmful to kiwi? It's because dogs have a powerful sense of smell, and kiwi have a powerful smell. Being creatures of the night, and having poor eyesight, they use a strong scent to identify themselves and their territories to others. Unfortunately, these odour trails can also lead a dog straight to a sleeping kiwi.

When confronted by a dog, a kiwi freezes. That response might have worked at a time when its only enemies were other birds. But these days, to freeze is to die. Even if a dog just wants to play, the outcome will be tragic. Not being a flier, the kiwi has a small breastbone protected by few muscles. Under the pressure of a dog bite, its chest crumples, and the bird soon dies.

The challenge for Northlanders is to keep kiwi and dogs apart. But how do you do that when one of the biggest thrills for any dog owner is to see their pooch run free? The dog/kiwi problem is really a people problem, and in Northland, people are fixing it. A few years ago, news reports and photographs of the mangled remains of kiwi that had been 'dogged' were common. Now, any news of dogged kiwi is much rarer; but if it does happen, the word spreads like wildfire through the community. Where did it happen, when, whose dog was it, whose kiwi was it?

What hasn't been realised until recently is how much the death of a kiwi can upset its surviving neighbours. Like us, kiwi live in communities and interact with their neighbours. A dogged kiwi will likely have a life partner. If one bird dies or disappears, its partner will spend each waking hour calling for its lost mate. This may go on for a week or more. If the lost partner does not return, the survivor will leave its territory and attempt to find it. This journey will take it through the territories of other kiwi, where its presence and its calls upset the whole community. One kiwi death affects many lives.

A specially trained dog (above top) follows the distinctive scent of a kiwi to lead a DOC ranger to a burrow, so that its owner can be fitted with a transmitter and monitored. However, dogs are a big problem for kiwi in many parts of New Zealand. The worst offenders are urban dogs such as the one that killed this Stewart Island kiwi (above).

The Backyard Kiwi team has discovered that a dead kiwi, such as a dog kill, leaves behind a grieving partner, who searches over many nights to find its mate, disrupting and upsetting all other kiwi in the area.

The 6000-hectare Whangarei Heads peninsula is kiwi country, and many people are buying into the idea that if we want to live here, it has to be on the kiwi's terms. New subdivisions are usually advertised as being cat and dog free. Farmers have always kept their dogs under control, and in the settlements, those who let their dogs run free — especially at night — are likely to get the hard word, not from DOC or the district council, but from neighbours. In fact, keeping kiwi safe from predators and dogs is a growth industry in many parts of Northland. There are upwards of 40 community groups putting in the hard yards on behalf of kiwi.

One such group is just over the hill to the north of the Backyard Kiwi mob of Whangarei Heads. Here, Mike Camm coordinates a group of kiwi-loving landholders, and with the help of a part-time trapper they are keeping over 3000 hectares predator free. The proof that it's working can be seen in a little bush-clad lake, where Mike rows his boat from time to

Just over the hill from the Backyard Kiwi communities of Whangarei Heads, Mike Camm coordinates another group of kiwi-loving landowners. They too are keeping their properties predator free (above) by trapping. As well as kiwi, rare brown teal (right) benefit from these initiatives. Once common in Northland, the teal are even more vulnerable to predation than kiwi. The 20 or more teal that come to Mike's pond now are a significant percentage of the Northland population.

time. When he does, he shares it with a flotilla of pateke, or brown teal. Brown teal were once one of the most common birds throughout lowland New Zealand. Now the 20 or so in Mike's pond represent about one-tenth of the entire Northland population; the only others are on Great Barrier Island.

The fact that they survive on this small lake is certain proof that predator trapping is working, because pateke are even more vulnerable to predation than kiwi. What makes these little ducks so vulnerable is their lifestyle. Mike demonstrates this point by tossing a handful of grain into long grass beside the lake. The little native ducks quickly come ashore and feed on the grain like pet chooks. They eat plants, grain, insects. Not only do they feed on open areas, they nest in the open too, which makes them sitting ducks for predators. But they are safe here, and hopefully in other places in Northland where community kiwi groups are out there regularly maintaining trap lines to catch those marauding stoats and wild cats.

Autumn is time for an annual ritual at Whangarei Heads. Friends and neighbours spread out over prominent hillsides for the annual kiwi call count. On this moonless night, down in the bush, kiwi are out and about and calling well. They continually advertise where they are to others in the area. Male calls are more falsetto and more frequent; from time to time, their partners will duet with a lower call. These calls allow counters on the hilltops to pinpoint pairs as well as single birds. The kiwi's screaming call is repeated up to 20 times and is easily heard from the nearby hills. When numbers are confirmed over a two-hour stretch on four different nights, and projected across the entire area, an estimate of total population can be made. From any listening site, this was somewhere between five and 16 in 2011. The total adult kiwi population is now estimated to be 400 at Whangarei Heads. But there are likely to be more, as there are known to be young birds, and young birds are silent. Even the thin piping calls of two-year-olds are unlikely to travel to the ears of wary kiwi counters.

Opposite

Soon after leaving the burrow each evening, a kiwi will call to re-establish contact with its neighbours. By monitoring these night-time calls, groups of kiwi counters from Backyard Kiwi can pinpoint where kiwi are resident and also make estimates of total kiwi numbers in different areas.

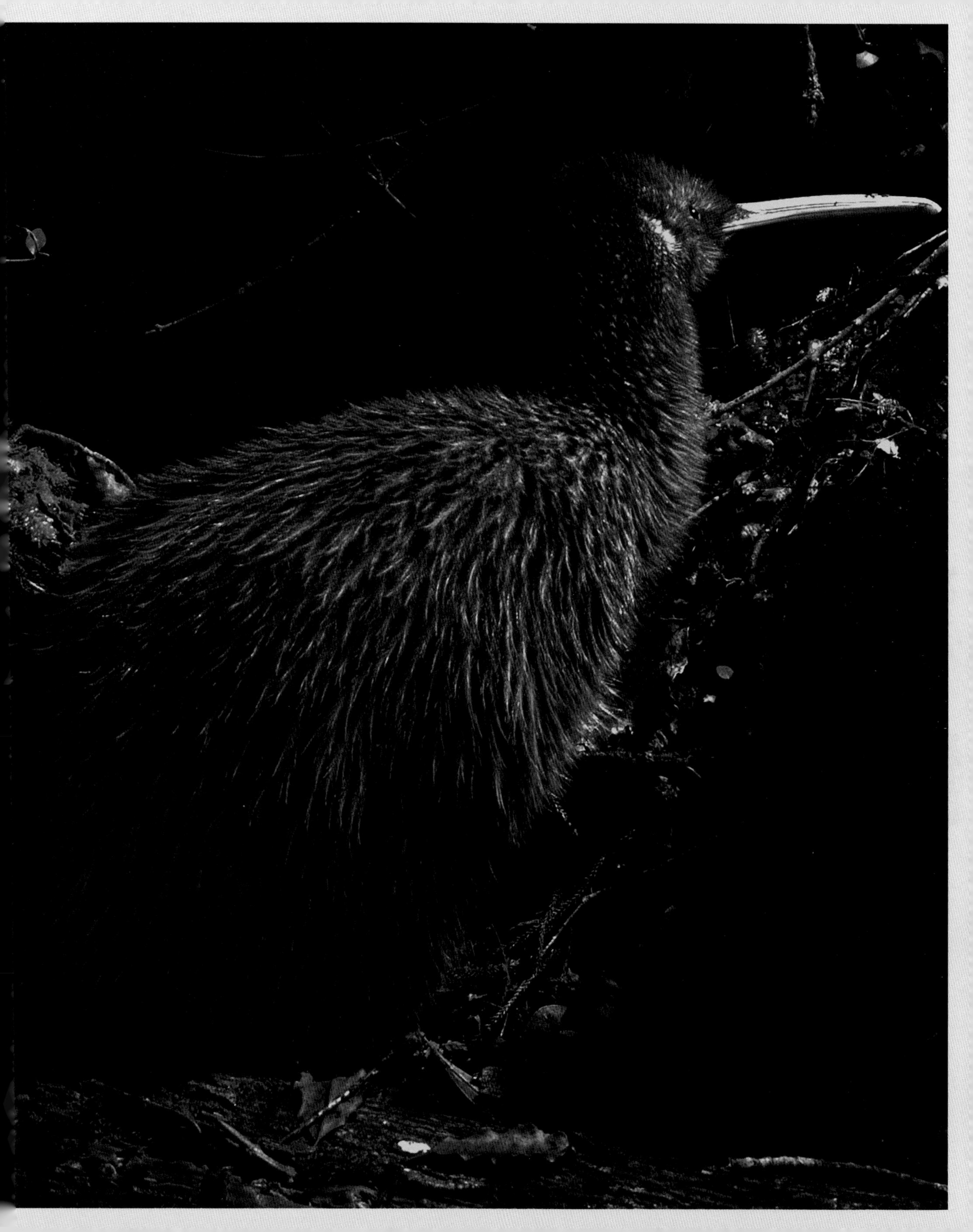

The moment has come for the kiwi chick named Christchurch. It has travelled by ute to the Onerahi slipway, then made a boat trip across to Matakohe/Limestone Island, a predator-free sanctuary in the middle of Whangarei Harbour. Now Todd Hamilton, island ranger Ben Barr and I head for a patch of bush along the north side of the island. We pass old Maori cultivations from which fernbirds take flight. We walk through a roofless, abandoned villa, the former home of the manager of the Golden Bay Cement company. Cement used to be made from limestone quarried on this 40-hectare island. But now it's a refuge for many rare and endangered species, including grey-faced petrels, several species of skink and gecko, fernbirds, banded rails, little blue penguins and young kiwi.

Now predator-free, Limestone Island in Whangarei Harbour is becoming home to a growing number of vulnerable reptile and insect species, as well as birds like the banded rail (above left) and grey-faced petrel (below left).

Full-time ranger Ben Barr (below) receives from Todd Hamilton of Backyard Kiwi the latest chick to be released onto this offshore refuge. The island was previously the centre of production for the Golden Bay Cement company, whose plant now lies deserted.

Ben has rigged up a temporary 'apartment' in a penguin box and Christchurch settles in as easily as if he'd been born here. Tonight he will emerge and begin the first stage of his new adventure. Like all other kiwi chicks that for the last 10 years

or so have been brought to this kiwi kibbutz, he will instantly adapt and grow quickly on the rich bounty of worms and other invertebrates that thrive here in the absence of rats and mice. Only when he is big enough, and reaches graduation weight of 1.2 kilograms, will he be taken back home.

Something quite special and wonderful is happening in Northland. Kiwis (the humans) are responding to kiwi (the birds) in a most delightful way. When word goes out that a chick is to be brought home from Limestone Island, it becomes a party. Hundreds of people gather at the rendezvous point, usually down a quiet country road. When Todd, or any of the other trained Northland kiwi handlers, arrives with the young bird, he becomes like the Pied Piper as mums and dads, kids and grandparents trail behind all the way to the release site. Everyone wants to be part of the event. Everyone wants to welcome their young kiwi back home again.

In a country where the national prognosis for some kiwi species is dire, with extinction a possibility in just a few decades, the work of Backyard Kiwi and other community groups gives us hope and inspiration that through people power we can increase kiwi numbers and keep them there.

In Northland, Kiwis (humans) are responding to kiwi (birds) in a remarkable way. Illustrator Heather Hunt has not only produced road signs and bumper stickers for Backyard Kiwi, her spellbinding images of kiwi also dance across the pages of a new children's book that she reads to a young fan (above right).

When a kiwi chick reaches 1.2 kilograms in weight, it is transferred from Limestone Island back into the community. Everyone gathers to welcome it home again. The kiwi handler, like the Pied Piper, leads mums, dads, kids and grandparents to the release site (below right).

beetle-battler

Chapter 12

A Window

What's the view like from your window? I have seen some pretty good views over the years, but none holds a candle to the view from the office of tour company Southern Discoveries in Milford Sound. It looks out, over John Robson's cluttered desk, to boats riding gently at anchor on soft grey waters that mirror a big, bold southern sky, but it's the bit in the middle that's the jaw dropper. Dead ahead is Mitre Peak. Today, the peak rises glistening from a tranquil Milford Sound. But don't be deceived, for tomorrow it might hide behind swirling angry clouds that glower above black, storm-lashed waters. Mitre Peak is the most photographed natural feature in New Zealand, but no photograph I've seen captures its sheer, defiant, lonely power. That power was forged in another age, an ice age.

Mitre Peak is a rock of ages: carved and etched, chipped and chopped by glaciers. Tens of thousands of years ago, it stood at the crash point of several intersecting valley glaciers that moved west, grinding, creaking and carving the sheer-sided fiord that draws and awes us.

I have never seen a view that compares with the one out of John Robson's office window in Milford Sound. Above the boats and below the sky is the towering icon of Mitre Peak. Its moody grandeur inspires all who see it. But it inspired John to go one step further and take steps to protect this threatened wilderness.

Just to the left of Mitre Peak is Sinbad Gully, a valley shaped by one of the glaciers that shoved at and shaped the peak. Now, above the valley's forested floor, its walls rise into the clouds like castle ramparts. Impressive. This is why I am in the office of John Robson, who has an important part in this story.

Over 40 years ago, a young man camping high in Sinbad Gully looked down and across the fiord to where I stand now. He had a bold idea about this valley, and that's what has brought me here. The man was Don Merton, 'the kakapo guy'. He was in the area as leader of a team searching the Sinbad and other valleys nearby for what they believed were the last kakapo left on mainland New Zealand. It seemed incredible that kakapo, once found from one end of this land to the other, could be on their last legs. If these valleys were their last refuge, what a magnificent but forlorn place for any creature to live or die.

Don's team had spent the previous night trying to prevent their tents from blowing away. And yet, strangely, it was storms like this, and winter avalanches, that buried the upper valley under tens of metres of snow, which had kept the flightless

In the 1970s helicopters were used to access the last remaining male kakapo living in remote Fiordland valleys like the Sinbad (below left). In summer Don Merton (below) and a team of kakapo rescuers observed that males puff themselves up and boom forlornly (right) to attract females that no longer existed here. He saw the Sinbad as a mainland island or refuge for endangered species, but it was an idea ahead of its time. He finally obtained permission to transfer some birds to offshore island sanctuaries, as Richard Henry (below right) had attempted in Dusky Sound over a hundred years earlier.

parrot safe. That, and the protection afforded by rock on three sides and ocean on the fourth. The fortress-like valley had given Don the idea that the Sinbad need not be a monument to the last kakapo, but a haven, a safe place, a mainland island sanctuary where kakapo and many other endangered species could live free and safe from predators.

The idea of Fiordland as a last refuge had been tried a hundred years earlier by a lone ranger called Richard Henry. He saw the predators approaching, and captured as many kakapo as he could and rowed them out to Resolution Island in Dusky Sound. Tragically, his island sanctuary was penetrated — because stoats can swim. A century later, predators were even beginning to penetrate Sinbad Gully. But Don argued that they could be controlled and that the natural defences of the Sinbad could be used to protect kakapo and other struggling species out there in our forests.

Forty years after Don Merton's exploits of the 1970s, I visited Sinbad Gully with Doug Keith (above) to follow up on some exciting new discoveries. By then, the old male kakapo like Butterbur (above left) were long gone, but his booming bowls (below left) still remained, like signs of a vanished community, on the ridge leading into the head basin.

I could see what he meant. We pushed our way through knee-high tussock, hebe and buttercups of the upper valley towards a 300-metre sheer rock wall. On each side, the walls were equally impressive. With a permanent snowfield above, they do indeed form an impenetrable fortress on three sides. Behind us, at the bottom of the valley, the sea completed the barrier. A breach of defences was all but impossible.

I was in the valley with my co-author, Rod Morris, who, back in the 1970s, was a member of Don's 'last round-up' team. He pointed out where a male kakapo named Butterbur 'boomed' out his hopeful love song across this starkly beautiful valley. This butter-yellow kakapo was people friendly, so a hide was set up to observe him. But the kakapo decided the hide was there for his amusement and would take a break from booming to use the roof as a slide and a jungle gym. He then began to use the space below the floorboards as a booming bowl.

Each night during the breeding season, Butterbur would make his way up from the beech forest to a few stunted trees on the narrowest and steepest spurs that had evaded being raked clean by winter avalanches. Precarious does not begin to describe his booming site, but it perfectly described his predicament.

Don had permission to take just six of the last 14 known kakapo out of Fiordland to safety on an offshore island. Unfortunately, Butterbur was not one of the chosen six, and as no one took up Don's idea of transforming Sinbad Gully into a predator-free sanctuary, it was to be where Butterbur died.

Back then, while searching for the last kakapo survivors, Rod discovered an unusual alpine gecko, the first ever seen in Fiordland. What was more remarkable was that it was active by day,

a trait not common among our geckos. Thirty years later, his discovery was confirmed by another gecko sighting, which came about in the most unusual circumstances.

In 2001, two climbers penetrated the Sinbad on a location-scouting mission for an upcoming climbing movie. When they spied the huge vertical wall at the head of the valley, they knew it was what they had been searching for. The wall would be perfect for the movie. It was slightly overhanging, dry, clean and, most important for climber comfort, it remained in shadow through the heat of the day. They named the wall Shadowland, a translation of Atawhenua, a Maori name for Fiordland.

Of course, the climber-scouts had to test themselves against the Shadowland wall. About halfway up, as one of them reached for a handhold, he came eye to eye with a short-tailed, beautifully patterned gecko. On their return, the scouts reported their discovery to DOC, along with some not so good news. Mice had got into the packs and food at their camp. Mice and geckos don't mix. The tiny rodents can squeeze into gecko crevices and prey on the occupants.

The Sinbad continues to offer refuge for animals that have long vanished from most of Fiordland. In 2001, high on the sheer Shadowland wall of the head basin (above), two rock climbers found a new alpine gecko (right) confined only to the vertical rock walls. In this extreme environment, searchers Tony Jewell and Trent Bell (below) found that the cold-blooded geckos maintain a comfortable temperature by hanging out on narrow vegetated ledges.

Late summer, the following year, Rod and two lizard experts returned to explore the wall in more detail. They favoured an area to the left of Shadowland, which had more sun, more waterfalls and more crevices and ledges for plant and animals to live in and on. Not far into his first climb, Tony Jewell discovered an unusual black-and-cream banded ground weta. It was huge, over 40 millimetres long. It was a new species, now named the superb weta. A short handhold away, he saw a large black skink. Again, none of them had ever seen anything like it.

Later expeditions discovered that a procession of wonderful creatures lived up on the wall. These included skinks; ground and cave weta in their thousands; and many unusual flies, moths, beetles, earthworms (as thick as your finger) and caterpillars, as well as spectacular slugs. The humidity from the waterfalls also made the wall a perfect home for thousands of olive-green, yellow-headed, leaf-patterned slugs. The scientists realised they were on a near vertical lost world of unmatched diversity. This was possible only because the wall was predator free, an almost unique situation in Fiordland.

Researchers discovered that the Shadowland head basin (above left) is a 'lost world' for large and spectacularly coloured reptiles and invertebrates. The sheer walls exclude introduced predators, creating a natural 'island of survival' for newly discovered species that include the Sinbad skink (top) and two new ground weta (right), one of them a giant (above right). The Sinbad peripatus (above) is also only found here.

It was once believed that bestowing Fiordland (above left) with not only National Park but also World Heritage status was enough to protect its biodiversity. But such status offers little protection against introduced predators. In the alpine zone the big threat is mice (below left and bottom). Unlike rats, mice are extremely cold tolerant and are active throughout winter, even on the coldest days. They are a significant threat to many plants, reptiles, insects and small birds like the rock wren (below). Herpetologist Tony Jewell and film-maker Jinty MacTavish felt something had to be done to protect the unique Sinbad Gully.

In 2004 Rod was back again, this time accompanied by two students making a film with herpetologist Tony Jewell, who knew the nature of this wall better than anyone else. At the end of a long day's filming, as they sat eating dinner on a rock looking down-valley, Tony remarked that something had to be done to protect this piece of Fiordland. He explained that the wall was a unique window offering a glimpse of ancient Fiordland, as it would have been before the heavy impact of predators.

When I heard this, I could not believe it. Surely, much of Fiordland National Park is predator free? Alas, no, the wall is all that's left. In fact, I felt I was looking at perhaps the last fragment of New Zealand that could be called original. But how do you protect a wall already protected inside a national park? It was film-maker Jinty MacTavish who offered the answer. 'We must tell people,' she said. 'No one will care until they know and can see for themselves what's living up here.' This led the three of them on a series of talks that inspired local people into setting up a trust to raise money that would protect the wall and its inhabitants. It hadn't happened a moment too soon. Mice and possums were closing in, moving up the valley.

When Butterbur the kakapo was booming out his forlorn love song in the hope that a female would dance and then mate with him, he could not have known that he was singing to ghosts. The females had all been killed. Because they nested on the ground, they were sitting ducks for stoats. But had the predators reached the upper basin beyond the tree line?

As we picked our way through boulders on the final stretch up to the wall, we were goaded on by the high-pitched peeps of rock wrens. Having an audience of several of these tiny bobbing birds was encouraging, as it meant there were few mice up here. If mice existed in big numbers, they would have wiped out the rock wrens, as they are small enough to get into the birds' burrows and eat the chicks.

In the distance, at the base of the wall, three figures waved to us. We joined them just after midday, by which time Shadowland was in shadow. Further along the wall, DOC scientists James Reardon, Hannah Edmonds and 'Jono' More were counting sunbathing geckos. They were trying to build up a better picture of where the lizards live and what they do. They estimated a population in the high hundreds or low thousands. They were also seeing lots of slugs and worms and weta.

The conversation turned to mice. The bad news was that they had made it onto the wall; the good news was that there were only a few and they were making little impact. However, James was concerned that if there were to be a major seeding year for beech trees in the valley, as happens every few summers, mouse numbers would explode. And having eaten all the seed, hungry hordes would move up the valley, likely bringing stoats in pursuit. It was a sobering scenario. Is this lost world about to be lost for good?

Thirty years after he took kakapo males out of Fiordland, Don Merton was reunited with the sole survivor, the bird named Richard Henry (above). Finally, his dream of turning the Sinbad into a mainland island is becoming a reality. A major Fiordland tour company supports DOC and the community in controlling predators in the Sinbad. These days, visitors to Milford Sound on Southern Discoveries vessels can feel proud that part of the price of their ticket helps fund conservation in Fiordland.

This brings us back down-valley, across the water, to the office with arguably the best view in the world. John Robson was asked by the trust if his company could help control predators in Sinbad Gully. He needed little persuading; he had long looked at the view, and he too shared the vision that by removing predators and building up native species, conservationists could turn it into a mainland island — one to which, one day, kakapo might return. As a first step to achieving Don Merton's dream from 30 years earlier, he successfully persuaded his company to fund predator control in the Sinbad.

These days, passengers on Southern Discoveries vessels pass close to the entrance of Sinbad Gully. As they gaze up to the deep-green forest and the cloud-topped wall beyond, their guide invites them to imagine a vision of a lost world where strange and unique birds, bats, reptiles and insects will live in vast numbers, just as they once did on the whole of this land. They learn that the price of their ticket is helping to make this vision a reality and that hopefully, on their next visit, they might get to set foot in this wonderful world of possibility.

It is not just people like John Robson and Don Merton who get swept up in the idea of returning parts of mainland New Zealand to how it was. It is an idea that has caught fire in the

save the remaining
The Sinbad Sanctuary Project
This project encompasses Sinbad Gully. Positioned below Milford Sound's iconic Mitre Peak, the gully is surrounded by cliff faces and is a natural fortress for rare and unusual wildlife. It was also the site of the last kākāpo found on mainland New Zealand.
sponsor, goes towards eradication programmes that protect native species like the whio (blue duck) that has inhabited this region for thousands of years.
The naming of Sinbad Gully.
ference.
southern discoveries

hearts and minds of men and women up and down the country, who believe in the possibility of recreating these lost worlds, and our obligation to this last place on Earth to feel the impact of humankind.

With our fire and our furry fiends, we completely clobbered the mostly mammal-free ecosystems of this delicately balanced bird world. In an unbelievably short time we have rendered one-third of our species extinct and another third endangered. If we wish to advance as a civilised people, we need to look at our record and agree that we want to turn it around. Private companies and communities throughout the country are helping by supporting DOC, and forming partnerships to repair the ecological damage in their areas. More than that, they are taking the lead. Mainland sanctuaries are a reality.

We are limited only by the extent of our self-belief. But we live in a nation of believers. The world-renowned rat catchers believed that offshore islands as big as Campbell Island could become rat free, and they were right. Next will come populated islands, such as Great Barrier Island and Stewart Island. It can be done. We don't want our most special species accessible only by climbers in the highest, most hostile and obscure corners of this land; we want them back where they belong — with us.

The Sinbad Gully is not only a window on the past, it is a metaphor for the whole country. During the advance of predators into this challenging environment, some populations and species went extinct; others retreated before the predators. Some, like kakapo, went to the brink and had to be rescued; while still others, like the cascade gecko and the superb weta and all the other crevice dwellers of Shadowland, stand firm, their backs to the wall, waiting — for what?

As I drove home after visiting the Sinbad, I paused on the Milford Road, beside the Gulliver River (above). I gazed up to survey kakapo Richard Henry's final home. It is sobering to realise how high kakapo had to climb to escape predators. But even that wasn't high enough, as it turned out. Perhaps sometime in the future, when all the predators have been eliminated from the Sinbad, the geckos, weta and other native creatures will come down and reoccupy the forests (right). Wouldn't it be wonderful if one day kakapo could join them?

Index

Images are indicated in bold.